MAKING SCIENCE FAIR

How Can We Achieve Equal Opportunity for Men and Women in Science?

Robert Leslie Fisher

University Press of America,® Inc.
Lanham · Boulder · New York · Toronto · Plymouth, UK

University Press of America,® Inc.
4501 Forbes Boulevard
Suite 200
Lanham, Maryland 20706
UPA Acquisitions Department (301) 459-3366

Estover Road
Plymouth PL6 7PY
United Kingdom

Printed in the United States of America
British Library Cataloging in Publication Information Available

Library of Congress Control Number: 2007929169
ISBN-13: 978-0-7618-3795-4 (paperback : alk. paper)
ISBN-10: 0-7618-3795-7 (paperback : alk. paper)

™ The paper used in this publication meets the minimum requirements of American National Standard for Information Sciences—Permanence of Paper for Printed Library Materials, ANSI Z39.48—1984

For Esther Schwartz Fisher (1910-1992);
Marie Ryan (1922-1999); and Shirley R. Anderson

Table of Contents

List of Tables

Preface

Robert Fisher
Delmar, NY
February 2007

This book has an unusual background. It began as an article submitted to a refereed journal. The editor turned it down but then asked me to expand it into a stand alone essay t he would consider publishing as part of a series of policy oriented works. When I finally completed the manuscript and submitted it, I learned the editor had passed away unexpectedly after surgery. I then cast about for a suitable publisher and, fortunately, found a sympathetic editor, Patti Belcher, at University Press of America. With her encouragement, I submitted the manuscript to University Press of America, who also published a previous work *The Research Productivity of Scientists* (2005.

Making Science Fair is both a continuation of the research reported in *The Research Productivity of Scientists* and a radical departure. On one hand, it looks more intensively at the issue of gender differences in reported need to alter or abandon research because of institutional controls than was possible in the *Research Productivity of Scientists.* For example, it emphasizes *disciplinary* differences in how strongly gender affects reported need to alter or abandon research because of institutional controls.

On the other hand, *Making Science Fair* it is a very different book from *The Research Productivity of Scientists. Making Science Fair* is a call to action, specifically, action and research to change the way scientific contributions are evaluated. The book asserts measuring productivity incorrectly is not only unfairly victimizing women and under represented minority STEM professionals, it is leading American universities to become dependent on foreign STEM professionals to staff their departments at a time when the availability of these foreign scientists is uncertain.

The President of the United States, George W. Bush, has taken notice of the problem of under utilized scientific research resources in his State of the Union message in early February 2006 citing the need to fund more scholarships and fellowships for STEM professions.

Making Science Fair is not merely a call for others to develop sound approaches to measuring productivity. It incorporates detailed proposals to accomplish this goal, especially an Appendix to "Making Science Fair" describing a tool for measuring research productivity. That paper, "Measuring Research Productivity in a Gender Neutral Way" grew out of a paper I presented with Carmela Triolo della Porta in November 2005 in the poster sessions of the Association of Public Policy Analysis and Management meetings in Washington D.C.. Dr. della Porta was very helpful in pointing me to the considerable public administration literature and emphasizing how developments there could be relevant to developing a gender neutral measure of research productivity to replace the unreliable publications count generally used in much writing by sociologists of science. However, Dr. della Porta in way bears any responsibility for the essay in this book.

Acknowledgements

This book is the second product of a long-term study of social factors in research problem choice that I began as a doctoral student at Columbia University and now am continuing since retiring from New York State service in 2003. Most of the people who have contributed to this study I have already acknowledged in my first book, *The Research* Productivity *of Scientists* published in 2005 also by University Press of America. However, many people have specifically contributed to the present monograph and I am pleased to recognize their contributions here.

First I would like to thank my late colleague Dr. Steven Miles Sacks, the editor and publisher of *Scipolicy* Dr. Sacks was chiefly responsible for my writing "Making Science Fair," the title essay in this collection. Without Dr. Sacks' enthusiasm for this project and his deep knowledge of the policy science literature that he selflessly shared this essay might not have been written. He had not read the version of the essay that appears here before his untimely passing and bears no responsibility for its content. However, his reading of earlier versions and his pointed criticisms of them played a huge role in shaping the final product. I also would like to thank three anonymous reviewers that Dr. Sacks provided to read an early draft. Their trenchant observations forced me to rethink and rewrite material I had assumed was already perfect. Writing is truly a humbling experience.

I am also grateful to Patti Belcher, acquisitions editor for University Press of America, who encouraged me to submit the manuscript to University Press of America for consideration after Dr. Sacks' sudden passing. I am not sure where I would have turned had Patti not offered to consider this work at that critical moment.

Dr. Carmela Triolo della Porta played an essential role in the drafting of an early version of "Assessing Research Productivity in a Gender Neutral Way" that appears here as an Appendix to "Making Science Fair." Dr. della Porta's deep knowledge of the public administration literature and her willingness to play "devil's advocate" sharpened our poster presentation at the November 2005 meetings of the Association for Public Policy Analysis and Management (APPAM) where an early version of the Appendix was presented.

Various people have helped in the research for and writing of this monograph besides my late colleague Dr. Stephen Miles Sacks. For help in acquiring data on women scientists in Europe and Japan I thank Laura Creswell of the Organization for Economic Cooperation and Development. I also benefited from assistance provided by the Association for Women in Government and the staff of the Bethlehem (N.Y.) Town Library.

My friend and colleague, Dr. Gregory Arluck, an econometrician, helped improve my understanding of economic terms and ideas as did my colleague Dr. Andrew Thomas Carswell of the University of Georgia. Dr. Arluck also looked over the technical Appendix on computing a productivity measure and made some suggestions for improving the document. Dr. Carswell read the manuscript from beginning to end suggesting numerous changes to enhance the clarity of the manuscript.

Others also helped improve the monograph. My brother Dr. Joel L. Fisher and my wife Shirley R. Anderson read and critiqued drafts of the monograph. I also thank my wife for her patience, understanding, and especially for her great baking that sustained me through the months I worked on the monograph. Furthermore, I benefited from professional journalistic tips offered by my friend and colleague Sherry O. Halbrook, editor of *The Communicator,* the official magazine of the New York State Public Employees Federation.

Special thanks also go to Joanne Ruppel for her expert work in readying this manuscript for publication.

Finally, whatever the merits of book, I know all too well that numerous errors remain despite countless hours I spent checking figures and text and the assistance of so many other people. I take full responsibility for those errors that remain and hope that the product justifies the support it has received from so many while I worked on it.

Introduction

American science is in a crisis or soon will be. The symptoms of the crisis are easy to describe. American supremacy in various important technologies is under challenge from foreign competitors; American universities are not graduating enough doctorates in the science, technology, engineering, and mathematics (STEM) disciplines to meet the growing demands of government, academic, and industry employers; and American universities, in particular, are recruiting foreign STEM professionals in large numbers to make up for the shortfall in homegrown STEM professionals.

According to the National Academy of Sciences, which studied the issue in 1991, the United States could train enough STEM professionals if it would recruit women and under represented minorities for these disciplines. However, American universities are reluctant to embrace this approach. Deep skepticism among the senior faculty members about the aptitude of women and minorities for technological subjects lies behind these academic institutions' unwillingness to add women and minorities to make up for the declining numbers of white males enrolling in STEM doctoral programs.[1]

Why should academic institutions hesitate to recruit women for the STEM disciplines? Today women are flocking to law schools and medical schools in record numbers. A woman sits on the Supreme Court of the United States, the highest court in the land; women fly the space shuttle and pilot commercial airliners; they design buildings and have even invaded the professional boxing ring.

Yet the doors to the professions in the United States have opened just a little since 1875 when Chief Justice C. J. Ryan of the Wisconsin State Supreme Court argued against admitting Lavinia Goodell to the Wisconsin Bar. "Nature," the Chief Justice declared, "has tempered woman as little for the judicial conflicts of the courtroom as for the physical conflicts of the battlefield." And, he continued, "Woman is modeled for gentler and better things. Our . . . profession has essentially and habitually to do with all that is selfish and extortionate, knavish and criminal, coarse and brutal, repulsive and obscene in human life."

In 1998, Virginia Valian summarizing a great deal of data on women in law, remarked "data from private law firms, corporations, and the judiciary" show "women are over represented in junior positions and under represented in senior positions." Furthermore, "Women advance more slowly than men do, even when their qualifications are equal" [and] "women's salaries rise more slowly than men's."

The world of the academy is similar to law and other professions. This is as true of science as practiced in the nation's universities as it is of the humanities. Science which prides itself on being a merit based system in fact is not. Women are a tiny fraction—about four percent—of the faculty of American physics departments and engineering departments. In no discipline are they much more than ten percent of the total of the senior professors in the scientific, technological, engineering, and mathematics (STEM) disciplines (Pinker [2005]; National academy of Sciences [2005]). "Women in academia are substantially under rewarded," Valian observed, pointing out "They are paid less" and take longer to receive tenure and promotions.

Are women under rewarded or are they rewarded properly? The uproar when Dr. Lawrence Summers, former President of Harvard University, told an audience of scientists in January 2005 women were not suited to do first-rate science, mainly because they would not make the necessary sacrifices to rise to that level of achievement and may not have the intellectual aptitude, suggests Dr. Summers voiced views quite outside the mainstream of American opinion. However, within the world of science and especially within the academic world, the opinions he stated are widely shared among senior faculty in the STEM fields. These views cannot be dismissed as merely the quaint ideas of a group of slightly senescent old men. They are an important element in the loud demands emanating from the academic sector that the United States Government must not put severe restrictions on entry into this country of foreign STEM professionals.

In the *short term* whether Dr. Summers is correct or not may not really matter from a policy standpoint. Skepticism about women's suitability for STEM professions has been ingrained within the American educational landscape since colleges and universities in the United States began offering science and engineering education more than a century ago. Consequently, the pool of women with the right academic credentials of outstanding publications, and prestige within their professions may not be large enough for Harvard to find more than few suitably qualified women to add to its star studded faculty.

The reality that there may be relatively few women for Harvard to tap is a prospect all thoughtful Americans should find alarming because the United States' economic and military strength depends on its scientific and engineering prowess. That preeminence is being eroded: fewer men are entering the sciences, as measured by doctoral degrees, and white males from whom the ranks of the nation's scientists and engineers have been filled are a declining proportion of the American population. Either American universities must recruit from women and under represented minorities to meet its needs in the future, or the United States will need to step up recruiting scientists and engineers from abroad, a source it has already tapped to make up for shortfalls in the numbers of qualified scientific and engineering (S & E) personnel from among American born people. Neither strategy is without some risks. Counting on foreign talent to overcome the deficiencies in American output of S and E professionals with doctorates is risky because the numbers of such persons willing and able to immigrate to the United States is highly unpredictable. Furthermore, tapping STEM professionals from troubled regions of the world raises security concerns. However, recruiting women and minorities will require changes in the way we train scientists and appraise their work.

Of the two policy choices, this essay takes the position the better approach is to look to women and minorities to assure an adequate supply of trained professionals in the STEM disciplines. The nation's political leaders agree as evidenced by President George W. Bush's State of the Union address in January 2006 calling for more support for STEM discipline training. Indeed, many prior works by prominent academicians and government scientists have made the same point (see e.g. Rosser [2004]; Hornig, ed. [2001], Etzkowitz et al [2000]; NAS [1991]). However, the present monograph takes a distinctive position in asserting foremost among the changes needed to assure an adequate supply of STEM professionals in the future is revamping the way we evaluate STEM professional productivity. The current methods of assessing scientific performance, in particular the publications count, are unreliable and biased against women. By fueling the false conclusion women are less suited to work in the STEM disciplines than men, they justify shortcomings throughout the educational system and workplace—especially failure to invest in training and promoting women and under represented minority STEM professionals. In short, unless we fundamentally rethink how we evaluate scientific performance we will never reach the goal of a sufficient supply of STEM professionals, including women and currently under represented minorities.

Regardless of how committed we are to do what it takes to achieve equality of opportunity for men and women, it will not happen soon. We are starting from a position where the ratio of men to women in most STEM fields is at least nine to one and in some as much as over twenty to one. Given no mandatory retirement age for faculty and longer life expectancies thanks to the startling advances in medical science, adding women to the top ranks of scientists and engineers is an ambitious long term goal and requiring a sustained commitment over many decades. We must make the commitment to achieving it, however, if we are to maintain our economic prosperity and military strength. That commitment must be embodied in changes we make now in the way we train STEM personnel and in how we appraise their contributions when deciding tenure and promotion.

The book's argument is laid out in the three following chapters and a Technical Appendix. In the first two chapters of this essay the focus is on answering the question: Why are women under represented in the ranks of American STEM professionals, especially at the top level? Is it because they are unsuited by attitude, interest, or intellectual ability? Or is it because of the hostile educational climate in which women train to become STEM professionals and a hostile work environment after women STEM professionals receive their degrees and embark on their careers?

This essay argues primarily a chilly educational and work climate has played a major role in women's lack of interest in careers in the STEM fields until the 1970s at least. However, over the past thirty years women's enrollment in graduate programs in the STEM fields has increased, though not uniformly. Women are concentrated in the social science and biological sciences programs. Few women enroll in graduate engineering or physics programs even today. And once women embark on their professional careers they still encounter discrimination, most of it subtle, but nonetheless damaging. They face difficulties in acquiring research support, higher teaching loads, and slower advancement in the academic sphere than men.

Universities are slowly facing the reality of their unequal treatment of men and women scientists. However, I assert that real progress toward equal opportunity for men and women in the STEM professions will not happen until the assessment tools we use, such as the publications count, are fundamentally changed. In Chapter Three, I present my detailed proposals for a new survey tool for measuring the gender gap in productivity. I also present my detailed proposals for individual scientist productivity assessment in the Technical Appendix.

Chapter Four summarizes the main argument and makes recommendations for changes in the milieus where scientists are trained and employed to make these environments more accommodating to women and under represented minorities. Many of the recommendations are similar to those already made by other scholars. However, this monograph also advocates developing new ways to evaluate scientific performance—a more controversial proposal. If this essay inspires work in that sphere it will have accomplished its purpose.

Endnotes for the introduction

1. See Pinker (2005). Also, see Becker (2005) for a statement of the position of the research universities. Rather than recruiting more American women and minorities, they would like to step up recruitment abroad of graduate students and scientists, and want easing of visa restrictions that were put in place after September 11, 2001.

Chapter One

Why are so few women in the STEM fields?

> We think it very important . . . there be no ceilings, other than ability itself, to intellectual ambition. We think it very important . . . every boy and girl shall know, if he shows . . . he has what it takes, the sky is the limit (Vannevar Bush 1945).

Do women have the aptitude to do first-rate science?

More than sixty years after Dr. Vannevar Bush declared science should be open to all people who have the ability and ambition to succeed, a view that is the official policy of the United States since 1945, women and many minorities (African Americans, Hispanic Americans, American Indians) are barely represented in some scientific and engineering disciplines and comprise no more than ten percent of senior faculty in most of them. Is this because white men have a greater aptitude than anyone else for science?

In January 2005 Dr. Lawrence Summers the then President of Harvard University, arguably the most elite institution of higher learning in the United States, suggested perhaps women are so little represented in the ranks of senior faculty at Harvard and other major research institutions because they would not make the necessary sacrifices and maybe they did not have the intellectual aptitude to do first rate science. Summers was roundly condemned in the mass media and by many faculty at his own university. However, he was voicing views widely shared by senior STEM professionals at his own university and elsewhere.[1]

Shortly after the public outcry erupted over Summers' remarks, a very interesting debate occurred between Steven Pinker and Elizabeth Spelke, two eminent Harvard psychology professors, in which Pinker gave a spirited defense of the Summers position. The full text of the debate, including Spelke's rejoinder, is available on the web by going to: *www.edge.org/3rd_culture/debate05_index.html.*

Before we turn to any other issue it is essential in my opinion to address the view men are somehow better suited than women to do scientific research, especially at the top level. First, these views about men's superiority enjoy support among many of the nation's intelligentsia and have important policy consequences such as the drive to recruit men STEM professionals abroad in preference to training more women STEM professionals domestically. Second, if these views are correct it seems pointless to discuss recruiting women for STEM positions. The entire purpose of bringing in women would be to buttress our nation's scientific and engineering prowess and if women cannot perform satisfactorily that purpose would be defeated.

Therefore, it is necessary to present Pinker's argument in depth about the superior suitability of men for scientific work and then critique it.

Pinker asserts the finest research institutions such as Harvard University draw their STEM senior faculty from the tiny fraction of the population who have not only high general intelligence and mathematical aptitude but also a set of other psychological characteristics found much more often in men than in even the most gifted women. While men and women have equal general intelligence they differ on other psychological characteristics some of which Pinker believes make men more likely to be highly successful scientists. Pinker identifies six characteristics men are likelier to possess than women that he maintains are relevant to scientific performance at the top level.

Pinker says, "The first difference, long noted by economists studying employment practices" is "men and women differ in what they state are their priorities in life." Pinker insists "men on the average are more likely to chase status at the expense of their families; women give a more balanced weighting" (2005:9). As support Pinker cites "a famous long-term study of mathematically precocious youth" who had been selected in the seventh grade for being in the top one percent on mathematical aptitude. Over the next two decades the researchers had re-interviewed this panel on attitudes and, according to Pinker, "there are statistical differences in what they say is important to them Pinker listed "ability to have a part-time career for a limited time in one's life" as an example of what women wanted. Women in the panel also desired to live "close to parents and relatives"; to have "a meaningful spiritual life"; and to have "strong friendships." Furthermore, women more than men, prefer to "work with people versus things."

According to Pinker, in contrast to women, men, desired "lots of money"; men also were interested in "inventing or creating something"; and finally they were more interested in "having a full-time career" and "being successful" in their chosen profession.

Pinker also emphasized more men reported they "would like to work 50, 60, even 70 hours a week." In other words, as Pinker nicely put it, "more men than women don't care whether they have a life."

A third difference is in the area of risky behavior. According to Pinker, men "are by far the more reckless sex." He cites a "large meta-analysis" in which "the largest sex differences were in 'intellectual risk taking' and 'participation in a risky experiment'" (2005:10).

Fourth, Pinker says a tremendous amount of psychometric data has accumulated supporting the idea "for some kinds of special ability, the advantage goes to women, but in 'mental rotation', 'spatial perception', and 'spatial visualization', the advantage goes to men." Pinker asserts many of the greatest physicists used "dynamic visual imagery" to hit upon their most famous discoveries.

Fifth, Pinker points to "mathematical reasoning" and asserts that researchers who studied mathematically precocious youth found that "the ratio of those scoring over 700 is 2.8 to 1 male to female." However, he concedes, "that's down from 25 years ago, when the ratio was 13-to-1" (2005:11).

Finally, Pinker points to the finding on an intelligence test administered in Scotland to the "entire population in a single year." The researchers found that "in the middle part of the range, females predominate; at both extremes, males slightly predominate." It is important to add that Pinker insists the six psychological differences he identifies are not simply the results of research in one culture, e.g. the United States, *but have been replicated in many societies.*

Now that I have presented Pinker's views, it is time to consider his position. I have one general observation to make and then I will consider specific points he has made.

Overall, I find, although Pinker makes many interesting observations, he lacks *appropriate* evidence for his conclusions. The question Pinker needs to address is, "Do a sufficient number of American women possess the necessary array of psychological traits and intellectual abilities to become first rate scientists?"

Pinker should be concerned with characteristics of a limited group of people —those with scientific aptitude. He needs to define this group *carefully* and then assess if women are equipped by nature with the requisite qualities to be included in it.

How does Pinker define scientific aptitude? Essentially, he equates scientific aptitude and mathematical aptitude. As evidence of their superior aptitude in science, Pinker asserts men predominate in the top ranks of mathematics ability on standardized tests.

Defining scientific aptitude as equivalent to mathematical aptitude is seductive but questionable. Without a doubt, mathematics plays an important role in all the STEM fields; however its importance varies across disciplines and even across specialties within disciplines. Put another way, mathematical skill is in varying degrees *necessary but not sufficient* for scientific achievement. It is highly important in theoretical physics to be a proficient, perhaps even an inspired, mathematician if one is to make significant intellectual contributions.

In many fields, however, while *some* mathematical proficiency is required it is also necessary to have other knowledge, skills, and abilities to do scientifically important research. The combined importance of these other specialized types of knowledge, skills, and abilities far outweighs the importance of mathematical proficiency in such disciplines as social science, biology, and archeology.

Consider the following famous example from low temperature physics. Some decades ago the conventional wisdom was that high temperature superconductors were not possible. Superconductivity was thought to be a phenomenon detectable only at near absolute zero (approximately –432 degrees Fahrenheit). This view received powerful support when three eminent physicists, Bardeen, Cooper, and Schrieffer formulated a rigorous theory that convinced nearly everyone high temperature superconductors simply did not exist and perhaps could not exist. Zuckerman (1978:84) remarked "the Bardeen-Cooper-Schrieffer theory of superconductivity was largely a culminating theory: 'After BCS theory, nothing else remained'." (Also see Moravcsik and Murugesan 1977:5).

As we all now know, however, some physicists refused to accept the Bardeen-Cooper, and Schrieffer theory. IBM, which understood the economic potential of high temperature superconductors backed empirical research on the subject challenging the theory by trying to find high temperature superconductors. Finally, a bench scientist in Switzerland succeeded in finding the first high temperature superconductor after working many years investigating different chemical compounds. The work he did was excellent science—he won the Nobel Prize for it—but it was not necessary for him to be an outstanding mathematician to do that work. The same can be said for many first rate contributions in biology, social and behavioral sciences, geology, archcology, and other fields.

Clearly, by equating mathematical virtuosity and scientific aptitude, Pinker shows how little impressed he is with empirical research of the sort the Swiss scientist did. He reserves his accolades for rigorous mathematically based theories

However, his personal preferences are not cogent grounds for accepting that men somehow are better equipped to do first-rate science. If he wishes to persuade us by force of logic, what he must do is look at what are the research problems in a field and what skills, knowledge, ability are necessary to solve them?

Besides his questionable equating of scientific and mathematical aptitude, Pinker makes other questionable assertions. Lacking evidence about the psychological traits of the component population of scientifically talented men and women, he cites *general* trait differences between men and women.

In general, if stringent statistical criteria are met, it is possible to make inferences about a larger population from a component population. This is the basis of survey sampling used in market research, evaluations of new drugs, and many other applications. However, it is never permissible to make inferences about a special component population from characteristics of a general population. Arguments that make these impermissible inferences are logically flawed because of what is known as an "ecological fallacy."

Pinker's attempt to demonstrate men with mathematical aptitude are superior to women with mathematical aptitude because men generally have characteristics more suited to science than women is an example of an argument based on an ecological fallacy. How relevant is the distribution of psychological traits in a large population to a special component of that population? In general, this is unknown. It is possible the elite group of men and women who have the necessary mathematical aptitude for science also have similar psychological traits, making them very different from the average men and women.

Now is the time to critique Pinker's position point by point. Recall Pinker indicates in a group of elite mathematically gifted individuals (those achieving scores of 700 or above on the Mathematics Aptitude Test of the SAT college entrance examinations when they took it in the seventh grade) he found various attitudes differentiated men from women.

One of the characteristics of the men with high mathematical ability differentiating them from comparably gifted women is their desire to make lots of money. However, this is an odd difference to point to for explaining why most Harvard STEM faculty are men. Relatively speaking, academic scientists do not make lots of money. It is more likely people who want to make lots of money would gravitate to business, or the profession of medicine or law, than to academic science.

Another characteristic Pinker points to is that women more than men value "strong friendships." It is difficult to see why women's interest in strong friendships prevents them from being good scientists. Is Pinker saying scientists work alone most of the time? If so, he is mistaken about much of science in the modern era. Much of science work today is team research. As Fox (1995:220) remarks, "Compared to the humanities, the sciences are more likely to be performed in teamwork than solo, to be carried out with costly equipment, to require funding, to be more interdependent enterprises Women's interest in strong friendship bonds would probably by advantageous in doing good science in team environments.

Pinker, however, might object that examining particular attitudes in isolation misses his point. Other traits differentiating men and women are relevant as well. In this vein, Pinker dives into a lengthy discussion of numerous other psychological traits generally differentiating men and women. However, Pinker *never* says these traits, e.g. "recklessness" or "spatial perception" or "mental rotation" etc., can discriminate between men and women of *high* mathematical ability (where such aptitude is treated as a reasonable proxy for people likely to be successful in the STEM disciplines generally). Does Pinker expect these psychological trait differences between men and women in the general population also are found in the component of highly mathematically gifted individuals—a small tail of a larger cohort of men and women? Pinker has problems no matter how he answers this question. If he does not have this expectation, his data are irrelevant. However, as indicated above, if he expects the differences in the general population also are present in the extremely mathematically gifted population, his argument suffers from an ecological fallacy—a serious flaw in logic occurring when one presumes what is true of a heterogeneous general population also is true of *any* subset in that general population.

Another serious flaw mars Pinker's argument: he apparently believes, from the standpoint of scientific achievement, where men differ from women the differences always favor men. Pinker might be inclined to cite the willingness to make mental leaps from the data, exhibited by Crick and Watson in their Nobel Prize winning work, as an example of a trait difference favoring men. Certainly, one can find many examples in the history of science of male scientists making mental leaps from their data. Examples are: the Alvarez hypothesis of a meteor causing the extinction of most dinosaurs and Einstein's work on relativity theory. An example of a female scientist who did the same thing as the two previous male scientists is Mary Schweitzer, who did work on soft tissue structures in the leg bone of a *Tyrannosaurus rex* (See Fields 2006).

However, most scientists who develop a theory by making successful mental leaps have benefited from the excellent empirical research done by prior investigators. Einstein had the benefit of Michaelson's excellent data while Crick and Watson could not have done their work without Rosalind Franklin's excellent spectrographic data.

Excellent empirical research also was necessary to test some of the elaborate theoretical ideas of physicists who made mental leaps. How would physicists know if there were any violations of symmetry until an experiment showed "left handedness" versus "right handedness" of particles? How would they know whether "charm" or "color" was the correct model of the subatomic particles without careful experimental analysis? Great science consists sometimes of work requiring mental leaps and sometimes of work requiring painstaking data collection and careful analysis of the results. If Pinker is merely dismissive of the high quality science work of Rosalind Franklin or Michaelson because they did not make the mental leaps that Einstein or Crick and Watson made, then his view of science is simplistic and uninformed about how science develops.

Pinker seems to think the willingness of men to make mental leaps somehow makes them better suited for science. A more informed view of science is that for its continued development it needs people willing to make hypotheses not firmly rooted in the available data and also people who test hypotheses carefully. If the latter is more characteristic of women – and some evidence (see Fisher 2005) shows women are more likely to be meticulous and cautious researchers—science would be better off with more equal proportions of men and women at Harvard and other renowned research centers!

It might be useful, certainly interesting, to close this discussion with some observations of Ben Barres, a neurobiologist at Stanford University. Barres began life as a girl named Barbara. Over a ten-year period of time when he was already past forty, he underwent gender reassignment surgery to become a man. In Barres' opinion, Pinker's views, which Barres calls the "Summers Hypothesis", amount to nothing more than "blaming the victim" (Begley 2006). As Barbara, Barres found he was discouraged in his quest to become a scientist. He reports that the "guidance counselor" at his high school advised him to go to a local college rather than apply to Massachusetts Institute of Technology (MIT). He applied anyway and was admitted.

While attending MIT as Barbara, he solved a particularly tough problem in a mathematics class that none of the male students could solve. His math professor "told me my boyfriend must have solved it for me. Furthermore, when he sought a sponsor for his thesis research, "nearly every lab head I asked refused to let me do my thesis research with him.

Barres devastatingly criticizes the tactic of trying to enhance women's status in science by publishing more papers. "I am still disappointed about the prestigious fellowship I lost to a male student when I was a Ph.D. student" (Begley 2006) he writes. The fellowship winner "had published one prominent paper" and Barres, as Barbara, had published six. (I discuss women's publication count and strategies to enhance it further below on page 41).

Who is selected for the top positions in science and engineering?

Simply because women have the necessary characteristics to be excellent scientists and engineers, but few choose STEM careers compared to men, is not necessarily a social problem. Two other facts make women's relative absence from the STEM disciplines a social problem.

One is the considerable evidence amassed that shows that women in their formative years have been steered away from careers in the STEM fields, or pushed out of such careers after beginning them. Those women who remain find their salaries and chances of receiving tenure and promotions are lower than men. The second is that all this occurs against a backdrop of academic, business, and even government recruiting of specialists abroad to fill positions in the STEM disciplines.

It is beyond the scope of this paper to debate whether foreign recruiting reflects a shortage in STEM professionals or simply an American employer preference for foreign professionals to the native talent (or at least the native talent willing to be recruited by the employer). It is possible to argue either side of the position as to whether there is a shortage of STEM professionals generally. Those skeptical any shortage exists can point out frequently engineers and scientists cannot find work because of changing priorities of major employers such as the Federal Government, industry employers, and the like. Cancellation of a major defense contract, for example, can result in massive unemployment in the ranks of particular kinds of STEM professionals. On the other hand, if there were no shortage why do we see the high wages and multiple job offers for freshly minted electrical engineers, chemical engineers, software developers and other highly trained specialists? These data testify to the fact the country has a voracious appetite for many kinds of highly trained STEM professionals not adequately fed by the available supply.

What is important to understand is the most prestigious scientists, the holders of endowed chairs and full professorships at the major research universities, overwhelmingly have doctoral degrees in their fields or related fields. The lopsided ratio of men to women in the senior ranks of American science and engineering certainly results partly from the near absence of women in the pool of eligible candidates, that is, persons granted the doctoral degree over the past forty years.

The prestigious positions in physics and other STEM fields in the United States are filled from not only American born holders of doctorates but also from foreign born and trained doctorates. The foreign born doctorates, even when American trained, are overwhelmingly men. An Indonesian woman physicist suggests why this is:

> in Indonesia, there is no different treatment between men and women physicists in their careers. The differences of opportunity for career development between men and women physicists are caused by natural characteristics and the biological function of women. In addition, women have more responsibility for household work compared with men. Therefore, generally, it is very difficult for women to achieve a doctoral degree or develop their careers (W. Subowo 2002:131).

Women are less likely to receive advanced education in most societies than are men because attitudes detrimental to women's obtaining advanced education are pervasive even in advanced western societies. The situation in Italy, the country that gave us Galileo, Leonardo da Vinci, Alexandre Volta, and Enrico Fermi among other giants of science, illustrates this point. Until the 1980s there were no doctoral programs in physics offered by Italian universities. Aspiring Italian men physicists would go abroad for doctorates in physics and then return home to find brides. In contrast, women generally marry at a fairly young age in Italy and once married cannot move abroad to continue their education. The option of going abroad was not as available to Italian women as to men and therefore, until the 1980s, most young Italian women could not really hope to obtain doctorates in physics. They needed to seek some other kind of degree they could obtain in an Italian university.[2]

Japanese women who receive advanced degrees outside Japan face prejudice from prospective mates and their parents. Many marriages even today are arranged between the parents of the prospective bride and groom. Japanese women who have gone abroad to get their college education or advanced degrees have found when they return home Japanese men and their families often will shun them as possible mates because of suspicion they do not have appropriate values (See Robert O. Blood 1967).

More recent research by Ogawa (2005) reports Japan has few women researchers in science and technology compared to other technologically advanced societies. According to Ogawa, compared to the United States or the United Kingdom, where 32.5 percent and 26 percent respectively of researchers in science and technology are women, only 11.6 percent of Japanese researchers in science and technology are women. Ogawa points to a variety of factors limiting the number of Japanese women researchers and reducing the likelihood of Japanese women reaching high positions in science and technology disciplines in the university setting. These factors include "heavier female care responsibilities" for children and the elderly than in western societies; "fewer job opportunities as a female researcher" and a "smaller number of female students majoring in the science and technology disciplines" (Ogawa 2005:6-10).

Japan is hardly unique in burdening women with heavier family care responsibilities than men. A common attitude in many societies is women will be mothers and wives and do not need advanced education, whereas men will have to earn a living and money should be spent to equip them to do this. In addition, women aspiring to train as chemists, physicists, engineers and other such professions encounter hostility among respected elders and men acquaintances to their seeking to advance themselves in a “man’s profession.”

The prevalence of the attitude women do not belong in a field such as physics is no doubt reflected in the numbers of women with advanced degrees in physics, chemistry and other STEM fields even in countries such as Germany. In Germany, according to Sandow and Kausch, the number of women earning a Ph. D. has risen from 96 to 152 between 1994 and 2003 (8% to 12% of all Ph.D.s). The number of female full professors rose from 19 (1.3% of all full professors of physics) in 1994 to 48 (3.6% of all full professors of physics) in 2003 (2005: 123). Women have to overcome not only attitudes hostile to their professional advancement, but also barriers in the form of prohibitive educational expenses for graduate training and little scholarship money or no information offered them about how to qualify.

These factors operate to some degree in the United States as well, though it is possible the problems have been worse in other societies, especially developing societies from which the United States has often recruited foreign STEM professionals. Regardless of the reasons, it is a fact more men than women receive Ph. D. degrees in the STEM fields in the United States. The National Science Foundation (See Appendix) reports of the 16,262 science and engineering doctorates awarded in 2001, 58 percent (9,395) were awarded to men. Obviously, the pool of male candidates for promotion to the senior ranks at the major research universities is larger than it is for women.

Reason for (cautious) optimism about the future

However, Fodor (2005:3) notes “it is not entirely for lack of qualified women, few women in the STEM fields reach high positions.” She quotes Elizabeth Ivey, president of the Association for Women in Science as observing, “There have been enough women in the life sciences pipeline for a number of years . . .

there should be more managers and department chairs and deans who are women." These impressions are corroborated by the numerical data (see Table 1) showing the pool of women from whom senior professors in science and engineering could be appointed is growing and has been growing at a faster rate than has the men's rate over the 35-year period 1966-2001. Since 1966 when just 924 women earned doctorates in the S and E disciplines, the number has steadily risen until peaking at 9,391 in 2000. In 2001, for the first time, there was a small decline to 9,319. Nevertheless, ten times as many women earned S and E doctorates in 2001 as in 1966. Eventually, some of these women, certainly more than in the past, will be candidates for appointment to the ranks of senior professors and may help redress the lopsided imbalance in the ratio of men to women full professors in the S and E disciplines.

Table 1: Doctoral Degrees Awarded to Women, by Major Field Group

Number of Doctoral Degrees Awarded to Women			
Academic Year Ending	**Total**	**Engineering and Physical Sciences**	**Earth, Math, Bio, Psych, and Social Sciences**
1966	924	128	796
1967	1,096	164	932
1968	1,317	187	1,130
1969	1,507	196	1,311
1970	1,648	243	1,405
1971	1,996	247	1,749
1972	2,151	271	1,880
1973	2,520	281	2,239
1974	2,671	268	2,403
1975	2,929	316	2,613
1976	3,097	299	2,798
1977	3,233	318	2,915
1978	3,454	300	3,154
1979	3,744	354	3,390
1980	3,961	412	3,549
1981	4,201	408	3,793
1982	4,350	481	3,869
1983	4,715	497	4,218
1984	4,792	550	4,242
1985	4,891	665	4,226
1986	5,167	735	4,432
1987	5,312	770	4,542
1988	5,662	853	4,809
1989	6,109	994	5,115
1990	6,371	1,076	5,295

Number of Doctoral Degrees Awarded to Women (continued)

Academic Year Ending	Total	Engineering and Physical Sciences	Earth, Math, Bio, Psych, and Social Sciences
1991	6,932	1,146	5,786
1992	7,080	1,276	5,804
1993	7,652	1,302	6,350
1994	7,921	1,463	6,458
1995	8,286	1,574	6,712
1996	8,649	1,619	7,030
1997	8,935	1,602	7,333
1998	9,348	1,700	7,648
1999	9,086	1,620	7,466
2000	9,391	1,672	7,719
2001	9,319	1,760	7,559

Source: National Science Foundation, Division of Science Resources Statistics, Survey of Earned Doctorates, 1966-2001.

In contrast, since 1990 the number of men receiving doctorates in the S and E disciplines has been declining after reaching a peak in 1995 of 12,082. It is reasonable to expect, therefore, the proportion of men holding the most prestigious professorships will drop in the future if only because the pool of candidates from whom to choose will have shrunk. However, that drop could be larger or smaller depending on whether the country embraces equal opportunity for women and men or continues to prefer recruiting foreign born and trained men scientists over native born women and under represented minorities.

Although overall women doctorates have increased and women have moved into the senior ranks in the academic world, women's rise in the ranks of the academic ladder has not been uniform across disciplines. For instance, women have made good progress in moving into senior positions in sociology at the top fifty U.S. research departments but much less progress in engineering or physical sciences according to Table 2, (see *Time* Magazine). Their results are consistent with observations made by scholars examining this problem for the past two decades (see Zuckerman 1987; Wajcman 1995; Mary Frank Fox 1995; Cockburn 1985).

Table 2: Doctoral Degrees Awarded in S&E and non-S&E fields, by Gender: 1966-2001

Number of Doctoral Degrees Awarded				
	S&E Women	**S&E Men**	**NonS&E Women**	**NonS&E Men**
1966	924	10,646	1,162	5,217
1967	1,096	12,013	1,346	5,948
1968	1,317	13,328	1,615	6,677
1969	1,507	14,781	1,881	7,574
1970	1,648	16,404	2,323	9,123
1971	1,996	17,385	2,600	9,886
1972	2,151	17,191	3,136	10,563
1973	2,520	16,853	3,565	10,817
1974	2,671	16,043	3,782	10,551
1975	2,929	15,870	4,272	9,881
1976	3,097	15,375	4,587	9,887
1977	3,233	14,775	4,625	9,083
1978	3,454	14,199	4,868	8,354
1979	3,744	14,128	5,193	8,174
1980	3,961	13, 814	5,447	7,798
1981	4,201	14, 056	5,691	7,408
1982	4,350	13,925	5,743	7,093
1983	4,715	13,920	5,818	6,828
1984	4,792	13,956	5,907	6,682
1985	4,891	14,044	5,853	6,509
1986	5,167	14,270	6,140	6,325
1987	5,312	14,582	6,120	6,356
1988	5,662	15,269	6,157	6,411
1989	6,109	15,623	6,404	6,191
1990	6,370	16,498	6,736	6,462
1991	6,932	16,985	6,941	6,539
1992	7,080	17,422	7,356	6,813
1993	7,652	17,568	7,469	6,816
1994	7,291	18,165	7,900	6,896
1995	8,287	18,117	8,129	7,044
1996	8,651	18,454	8,306	6,832
1997	8,936	18,089	8,313	6,862
1998	9,347	17,816	8,502	6,825
1999	9,086	16,737	8,398	6,698
2000	9,384	16,512	8,730	6,643
2001	9,303	16,162	8,598	6,607

Source: National Science Foundation, Division of Science Resources Statistics, Survey of Earned Doctorates, 1966-2001.

The United States has one of the lowest proportions of women scientists in the natural sciences, especially in physics. In 1991, Megaw (quoted in NAS, 1991) surveyed physics departments of the world and reported the United States and Korea had the lowest proportion of women faculty (three percent) in physics of any nation responding, followed closely by the United Kingdom (four percent) and Japan, Netherlands, and New Zealand (six percent). In contrast, Hungary reported 47 percent of its physics faculty was women; the Union of Soviet Socialist Republics (now Russia) reported 30 percent of its physics faculty members were women; and France, Italy, and Turkey were tied at 23 percent (Table 3).

Table 3: Doctoral Degrees in Physics and Women Faculty in Physics Departments (1991)

Country	Doctoral Degrees To Recent Women Grads	Women Faculty in Physics Departments
United States and Selected Western Nations	**Range: 4-29 percent of all physics doctorates**	**Range: 3-23 percent of all faculty**
United States	9%	3%
United Kingdom	12%	4%
Netherlands	4%	6%
Ireland	20%	7%
Belgium	29%	11%
Spain	21%	16%
France	21%	23%
Italy	21%	23%
Selected Asian & other Non-western Eastern European Nations	**Range: 4-60 percent of all physics doctorates**	**Range: 3-47 percent of all faculty**
Korea	5%	3%
Japan	4%	6%
India	26%	10%
Poland	17%	17%
Brazil	31%	18%
Turkey	17%	23%
USSR (now Russian Federation)	25%	30%
Philippines	60%	31%
Hungary	27%	47%
Adapted from Table 5 in NAS (1991). Original Source: W.J. Megaw, "Gender Distribution in the World's Physics Departments" paper prepared for the meeting, "Gender and Sciences and Technology" Melbourne, Australia, July 14-18, 1991.		

Proportionally, the representation of women in physics in the United States has not increased much since 1991. However, the United States is certainly not alone in having a miniscule proportion of women scientists, especially in the higher ranks of scientists. While up-to-date data comparable to Megaw's on the gender distribution of faculty are not available for every country, data for Japan and many European countries in the period 1998-2002 show a gender gap in full professors and researchers that is similar to the gap in the United States (see Appendix Table). As in the United States, there is a small discrepancy between women and men students in the sciences that widens as one goes from student to junior faculty to senior faculty for every country surveyed.[3]

These findings are a sobering reminder the United States and other technologically advanced countries have a long way to go in providing opportunities for women in the STEM fields within the academic sector. Will women have more success in penetrating the top ranks of engineering and physics than they have in the past? Before we can begin to answer this question we need to know whether the pool of women candidates for senior positions is growing and how many there will be for such positions in the future.

To answer these questions, the data in Table 1 are not ideal. A better idea of the candidate pool for top positions would be data on the doctorates in each discipline rather than aggregated data as in Table 1. Although not shown in Table 1, the National Science Foundation (NSF) reported the number of women with doctoral degrees broken down by discipline for the period 1966 through 2001. According to the NSF, in 1966 just eight women received doctorates in engineering. Even as recently as 1981 less than 100 women earned doctorates in engineering. However, the numbers have climbed considerably beginning in the mid 1980s. In 1990, for example, 415 women received a doctorate in engineering and in 1995, 696 women received their doctorates in engineering. In 2000 the number had risen to 837 and the following year it had reached a high of 927. To provide some perspective, the number of engineering doctorates awarded to women in 2001 is more than 100 times that awarded to women in 1966.

These numbers, however, exaggerate the pool of women engineers potentially available for senior positions in academia, government and industry because they include women who came to the United States to earn an advanced degree and who will leave afterwards, probably to return to their native lands.

The actual pool is more accurately reflected in numbers of women who are U. S. citizens and permanent residents. As the data in Table 4 indicate these numbers are much lower. For instance in 1994, according to Table 4, American institutions of higher learning awarded a total of 3,053 engineering doctorates to U. S. citizens and permanent residents of which women received 448, about 15 percent total. By 2001, the total number of engineering doctorates had dropped from its peak of 3,389 in 1996 to 2,435 of which women had collected 19 percent (474).

Table 4: Engineering Doctorates Awarded to U.S. Citizens and Permanent Residents, by Sex and Race/Ethnicity: 1994-2001

Engineering Doctorates Awarded to U.S. Citizens and Perm. Residents								
	1994	**1995**	**1996**	**1997**	**1998**	**1999**	**2000**	**2001**
Engineering	3,053	3,342	3,389	3,332	3,046	2,888	2,569	2,435
Female	448	520	526	512	452	492	490	474
Male	2,604	2,822	2,863	2,819	2,588	2,396	2,079	1,961
Unknown	1	0	0	1	6	0	0	0
White	2,020	2,092	2,263	2,287	2,170	2,110	1,866	1,746
Female	305	321	348	332	307	352	333	317
Male	1,715	1,771	1,915	1,955	1,863	1,758	1,553	1,429
Asian/Pacific Islander	865	1,032	896	707	554	512	439	422
Female	119	170	134	119	97	84	109	96
Male	745	862	762	588	454	428	330	326
Unknown	1	0	0	0	3	0	0	0
Black	54	70	74	98	83	98	81	92
Female	13	15	19	23	22	24	24	25
Male	41	55	55	75	61	74	57	67
Hispanic	66	77	99	97	110	82	82	91
Female	6	11	15	23	13	23	13	22
Male	60	66	84	74	97	59	69	69
Am. Indian / Alaskan Nat.	6	9	14	17	13	12	8	7
Female	2	0	2	3	4	2	4	2
Male	4	9	12	14	9	10	4	5
Other or unk race/ethnicity	42	62	43	126	116	74	73	77
Female	3	3	8	12	9	7	7	12
Male	39	59	35	113	104	67	66	65
Unknown	0	0	0	1	3	0	0	0
Source: National Science Foundation, Division of Resource Statistics, Survey of Earned Doctorates, 1994-2001.								

Considering the tiny fraction of female faculty in engineering schools, (perhaps as few as 3 percent), that women earned about 19-20 percent of the doctorates in 2000 and 2001 augurs well for substantial numbers of women moving into the top ranks of engineering in the future. However, based on the usual trajectory of academic careers, it will be perhaps twenty years before substantial numbers of the recent cohorts of engineering doctorates appear in the higher prestige positions within the engineering profession. Even then, men will probably be 85-90 percent of the senior engineering faculty, only a modest decline from over 95 percent currently.

Women have made progress in physical sciences too. In 1966, according to the NSF, 120 women received their doctorates in physical sciences; in 1970 it rose to 227; by 1980 it had increased to 322. It continued to increase rising to 467 in 1985; to 661 in 1990; and to 878 in 1995. The number of doctorates in physical sciences awarded to women peaked at 926 in 1998 and then dropped by nearly a hundred from 1999 to 2001 when it varied between 831 and 835.

The data presented thus far demonstrate men overwhelmingly fill the top ranks of their respective S and E disciplines in part because more men are available from whom to select. This, of course, only raises more questions. Why are women not found in the engineering doctoral programs in numbers comparable to men? Why did the number of women engineering doctorates increase dramatically beginning in the mid 1980s? Why did the total number of doctorates in engineering and other S and E disciplines grow steadily over the first 25-30 years of the 35-year span and then begin to drop?

Why are low numbers of women found in engineering compared to other disciplines?

Traditionally women have not entered engineering in significant numbers. Whereas women have made great strides in such fields as medicine, social sciences, they are proportionally a small fraction of engineers. In 1995, Mary Frank Fox noted "Men are . . . seven times more likely to be in engineering" than women in the United States. About 19 percent of men doctorates and only 2.5 percent of women doctorates were in engineering at that time according to Fox.

Feminist scholars have noted the small proportions of women in engineering arguing the engineering disciplines traditionally have discouraged women from seeking careers in that group of fields. Judy Wajcman (1995:202), for example, said, "Of all the major professions, engineering contains the smallest proportion of females and projects a heavily masculine image hostile to women As a result, there is a shortage of women engineer role models to act as a buffering influence and mentor other women into the field.

Wajcman adds, based on Hacker (1981) "engineers attach most value to scientific abstraction and technical competence" while Cockburn observes "in order to fortify their identification with physical engineering, men dismiss the intellectual world as 'soft' Wajcman further alleges, "The posing of such categories as 'hard/soft' and 'reason/emotion' as opposites is used to legitimate female exclusion from the world of engineering" (1995:202-203). However, as Cockburn points out men "need to appropriate sedentary, intellectual engineering work for masculinity too" (1985:190, quoted in Wajcman 1995:203).

If Cockburn and Wajcman are correct, the rationalizations that have been used to limit women's entry into engineering are little different from those once offered to block women from becoming lawyers. Thus in 1875 Chief Justice C.J.

Ryan of the Wisconsin State Supreme Court argued against admitting Lavinia Goodell's application to be admitted to the Wisconsin bar by observing

> Nature has tempered woman as little for the judicial conflicts of the courtroom as for the physical conflicts of the battlefield. Woman is modeled for gentler and better things. Our . . . profession has essentially and habitually to do with all that is selfish and extortionate, knavish and criminal, coarse and brutal, repulsive and obscene in human life. It would be revolting to all female sense of innocence and sanctity of their sex . . . and faith in woman on which hinge all the better affections and humanities of life (Epstein, 1970: 245, quoted in Sachs and Wilson 1978:96-97).

There is ample evidence women have not entered engineering and they were not obtaining advanced degrees in engineering in significant numbers until recently. How much of that is explained by women not finding the profession attractive and how much is accounted by hostility to women who sought engineering educations and positions is difficult to say with the data in hand. Anecdotal reports of women being told they do not belong in engineering abound, but hard evidence of women being systematically excluded does not.

Why did women quite suddenly begin to flock to engineering doctorate programs in the past twenty years? Most likely, the total numbers of doctorates in engineering grew in response to demand particularly from industry and also from academic institutions (albeit to a lesser extent) for more engineers and more highly specialized talent. Demand from the academic sector followed its growth in capacity (i.e., numbers of colleges and institutions of higher education and spaces in existing colleges to accommodate the huge numbers of students from the baby boom generation desiring space in post secondary schools, especially bachelor's degree granting institutions) which itself creates a demand for new departments, new terminal degrees, and hence for new staff with new credentials to teach the courses for the new terminal degrees. Finally, beginning perhaps in the 1980s, institutions of higher education in the engineering disciplines may have stopped up their efforts to recruit women into engineering, spurred no doubt by prodding from government and women's rights groups among others to open up spaces for promising women students.

Engineering disciplines may be the most resistive of all STEM disciplines to women's advancement professionally but in no STEM field can we find equality of opportunity for men and women. This situation is not because women lacked ability to do science but because of historical factors, that even today exert an influence on how universities and colleges behave.

Historical factors behind women's low participation in science and engineering

Women's low participation in science and engineering historically is a result of a confluence of social factors. The European societies from which most Americans originally came in the eighteenth and nineteenth centuries barred women from many occupations, especially professions such as medicine and law. These barriers continued when the United States gained its independence from Great Britain in 1783 and did not fall until the latter part of the nineteenth century or even the early twentieth century. Not only did women face barriers to their entry into certain occupations, women did not enjoy the same access to higher education men did until long after the first men's colleges were founded in the seventeenth and eighteenth century.

These barriers reflected the impact of traditional mores and ancient legal codes, e.g. the Roman legal code, which made women and children property of the father before marriage and the husband afterwards. Women's inferior status in law was manifested in many ways besides limited educational opportunities e.g. lack of the right to vote (not won until after World War 1) and even in lack of the right to custody of their children after divorce.

While legal barriers to women's advancement in the professions have largely disappeared, the social attitudes these legal barriers incorporated have not. The world of science and engineering, a world in which men are overwhelmingly dominant in numbers, still is not welcoming to women. Below I discuss four major historical and structural factors accounting for the chilly climate women have endured and that is likely to persist for some time in the future.

1. University settings as historically male preserves

We think of scientific research as highly technical work occurring in particular settings such as university laboratories and carried out by specially trained people with advanced degrees. However, this has not always been true. In fact, according to Etzkowitz, Kemelgor, and Uzzi, (2000:18), "in the eighteenth century, many laboratories, especially in chemistry, had been in kitchens in the home." With the rise of universities in the nineteenth century, Epstein remarks, we find scientific research work was henceforth mostly limited "to those having access to such locales (as the university laboratory)" (1992:241). She further points out, "Restriction of location, while permitting concentration of tasks and resources, also restricts the pool of scientific personnel to the like-minded and similarly experienced" (1992:241).

When science was done in kitchens, as Etzkowitz et al. (2000) observe, women, especially wellborn women, could participate freely along with men. However, when science work moved to the university setting, women were excluded. Why did this happen? And, more importantly, why has this situation persisted to the present time?

Part of the answer is American universities, which perform most of the scientific research in this country, have taken as their models their German, and (to a lesser extent) British counterparts both of which traditionally excluded women.

The first American universities such as Harvard and Columbia developed out of colleges that from their beginnings in the seventeenth and eighteenth centuries were all- male bastions. When women could go to college at the outset they could only attend women's colleges such as Barnard, Radcliffe, and Vassar. Even there, the faculty was often overwhelmingly, if not totally male, until well into the twentieth century.

Up until the beginning of the twentieth century, German universities forbade women from matriculating and women could only audit classes. The exclusion of women from German and American universities was relaxed over several decades in the twentieth century. For instance in 1900 and 1908, Baden and Prussia, two of Germany's states, permitted women students to matriculate for the first time. American universities also were permitting women to matriculate by the beginning of the twentieth century.

Epstein points to a 42 percent female enrollment at Tufts University in 1900 and to 26 percent of the graduating class of Boston University in 1902. (1970:248). However, she reports Harvard waited until 1945 to admit women to its medical school; until 1950 to admit women to its law school and until 1963 to admit women to its business school.

2. Occupations seeking professional status

The difficulties women had gaining entry to graduate schools of any type were not simply the result of quaint views of women as too tender to handle the stresses of professional training. Various occupational groups starting in the nineteenth century sought to enhance their economic and political status by obtaining professional status. For example, the leaders of these occupations sought to require all persons joining these occupations be licensed and/or meet educational standards such as a master's degree. The leaders of these occupation groups also realized as long as women could not vote—a right they gained only in 1919—fields with high proportions of women were less likely to gain recognition as professions and this could be a threat to their economic well being.

The professionalizing of an occupation need not erect barriers to women's entry. However, in the United States and most Western societies the drive to professionalize occupations often also went hand in hand with erecting barriers to women's entry into these occupations. These societies were heavily influenced by practices of the Judeo-Christian religions, which incorporate many barriers to women's holding high status roles. Women cannot be rabbis in the Orthodox Jewish sect. In Roman Catholicism, women cannot be priests or any other clergy. Women who choose the religious vocation have special orders and titles such as "Mother Superior" reserved for them and only supervise other women.

Nursing is an interesting case illustrating how a field with large numbers of women continues to suffer overall economic disadvantage in American society. Nursing, "a male occupation well into the nineteenth century had become a largely female field not long into the twentieth century" according to Etzkowitz et al. (2000:23). Nursing salaries have lagged until recently in part because of the field's image as a female occupation, and the belief many nurses would leave the field to raise children rather than fight for higher wages and better working conditions. However, nursing now has become more attractive to both genders, thanks to a shortage of skilled nurses, and concomitantly rising salaries at leading hospitals, particularly for specialized nurses such as nurse anesthetists and nurse practitioners.

3. Influx of foreign male students into STEM training programs in the United States

The growth in demand for scientists in the United States has outstripped the ability of American universities to supply qualified scientists unless they take action to improve recruitment and retention. American universities thus far have adapted to this demand for qualified scientific and engineering personnel by a combination of expanding their recruitment and retention of women graduate students and by recruiting substantial numbers of foreign graduate students.

Whereas a need to recruit women students might be expected to lead to changes in the culture of the university towards a more women–friendly climate, this has not always happened. Part of the reason is many of the foreign students recruited come from cultures where women are expected to be deferential to men and do not enjoy the same legal status and men do. The National Academy of Sciences (NAS) reported "faculty, teaching assistants, and graduate students from certain cultures are less accustomed to the presence of female students in the classroom and laboratory and may discriminate against women students either consciously or unconsciously" (2005:47).

Table 5: Graduate S&E Enrollment by Citizenship and Sex, and Postdocs in S&E Fields

	U.S. Citizens and Permanent Residents			Temporary Visa Holders		
	S&E Grad Students		S&E Postdocs	S&E Grad Students		S&E Postdocs
	Male	Female	(M & F)	Male	Female	(M & F)
1993	NA	NA	11,635	NA	NA	13,030
1994	195,794	133,232	12,469	76,237	25,879	13,318
1995	189,915	134,047	12,823	72,341	26,163	13,337
1996	182,519	134,556	12,930	70,991	27,115	13,639
1997	174,934	133,734	12,835	70,685	28,277	14,429
1998	169,490	133,389	12,966	71,939	30,038	14,910
1999	165,823	135,431	12,725	76,963	32,965	16,255
2000	157,023	133,688	12,627	86,034	36,791	17,597
2001	158,015	136,696	12,086	93,797	40,734	18,106
2002	165,004	144,276	13,523	101,244	44,368	18,381
2003	175,027	152,305	13,590	101,063	45,808	20,095

Source: National Science Foundation/Division of Science Resource Statistics, Survey of Graduate Students and Postdoctorates in Science and Engineering

Rosser (2004) recounts the tale of a successful computer scientist (herself a Polish immigrant), Dr. Marina Titelinska. According to Rosser, Titelinska moved with her husband to a "State Research 1 university in the Northeast" after he received an appointment there following completion of his Ph.D. degree studies. She received her MS. in computer sciences at the school and "began work on her Ph.D. She had an unpleasant experience with her advisor who came from Iran and for whom she was his first Ph. D. student." Rosser explains, "Because he wanted to control her, they fought over many issues, including publication of papers." Despite this bad experience, Titelinska managed to finish her doctoral studies at a "prestigious public university in a neighboring state and now has a fine position, in part thanks to her meeting a woman academic executive who supported and encouraged her" (2004:16). Both Etzkowitz and NAS, however, suggest such happy endings are not the rule in cases where women clash with foreign-born male authority figures.

After the U.S. embarked on a war on terrorism in the wake of the terrorist incidents of September 11, 2001, one of the steps taken to protect the security of the United States has been to scrutinize foreign nationals more carefully. It is simply much more difficult for foreign-born individuals, especially from developing nations, to enter the United States to work or study here. However, the universities are bitterly contesting the new restrictions on foreign talent. They have been campaigning vigorously in the media (see the remarks of Nobelist Gary S. Becker below) and the halls of Congress to be able to continue tapping this resource even though for many reasons we can expect a diminution of the supply of such immigrant talent in the future.[4]

It should now be clear that when science moved in the nineteenth century from the kitchen and garage to the university laboratory, women lost much of their opportunity to participate. Not only did the university setting resemble a male preserve, but also various societal forces, such as the drives to gain professional status for various occupations and to recruit foreign students, often from cultures where women are legally and socially subservient to men, contributed to making the university an unfriendly environment for women.

It is important to understand this unfriendly environment for women was not an ideal environment for men either. While it has met men's needs for prestige rankings and supplied qualified personnel to the workforce, it has also interfered with men's ability to be fathers to their children. Therefore, despite its mistreatment of women and minorities, as long as any shortfall in the economy's needs for scientific and engineering personnel could be met by importing foreign personnel, the research universities felt no compelling need to change the system. It is only now, with foreign scientists no longer readily available to make up for any shortfall, the need has become *urgent* for the culture of the university to be retooled.

4. The dysfunctional university culture

Before we can prescribe how to change the university culture, we must understand what the problem is. It is not enough to simply say it is dysfunctional and needs to be made more accepting of women.

The university culture resulting from the interaction of these three factors can be described in Etzkowitz et al. as the "weed-out system." They said, "In large universities at the bachelor's or first degree level, women often encounter a 'weed-out' system of courses based upon a competitive model that is designed to eliminate unwanted numbers of prospective students" (2000:49). And, they added, "This system has even worse effects on women than it does on men. Its encoded meanings, obscure to young women whose education was grounded in a different system of values, produce feelings of rejection, discouragement, and lowered self-confidence" (2000:49).

The weed-out culture is a hallmark of the large undergraduate colleges that have relatively less selective admissions policies. It does not exist in women's colleges and smaller elite colleges to anywhere near the same extent, according to Etzkowitz et al. (2000). However, many women, encouraged in their interest in science in such nurturing environments as the women's college or small elite college, as well as those women who went to large undergraduate colleges and managed to survive "this perilous journey" through the bachelor's degree process, "once in graduate school . . . often encounter a second weed-out system, a harsher, more discouraging version of the research model they experienced as undergraduates" (2000:50).

Etzkowitz et al. (2000) elucidated the "U.S. graduate education model" to show how the weed-out system came to be perversely anti-women. Although the weed-out system operates in graduate school for men students as well as women, according to Etzkowitz et al. (2000), for men it is ameliorated to a considerable extent by an informal culture. Etzkowitz et al. say "male professors draw many of their male graduate students into a supportive, caring environment. Such coteries surrounding a faculty member, typically including students from many nations and cultures, constitute the basic social unit of U.S. doctoral education" (2000:72). The bonds forged in this social unit are strong enough to overcome the suspicions and hostilities students from countries that are at odds with each other might feel for one another ordinarily. Indeed, it is expected the members of this circle will form bonds that "will last a career, if not a life time. Depending on the discipline, the coterie may consist of students working in the same laboratory (experimental sciences) or from members of the same seminar group (theoretical sciences)" (2000:72).

When there are few women students this informal culture within the graduate school department may exclude all of them (Etzkowitz et al. 2000). "Contrary to gender stereotypes, female graduate students are often left to be the 'rugged individualists', having to fend for themselves" (Etzkowitz et al. 2000:74). Etzkowitz adds the "precarious status" of women graduate students takes a toll in "predictable social and psychological consequences that, if not countered, eventually affects [their] scientific work" (2000:**). He quotes a woman graduate student, "I felt like I didn't have any back-up support. I didn't know how to pick a topic. The guys talk about that at the bars. I don't go there" (2000:74). Etzkowitz et al. points out although informal study groups are important to success in graduate school, "We identified some degree of exclusion [of women students] from study groups in virtually all departments studied" (2000:74).

Despite the fact we are now in an era when universities should be trying to increase their science and engineering graduates, both male and female, the "weed-out" system persists. As Etzkowitz et al. point out, the weed-out system is "a long-established tradition in a number of academic disciplines" and especially "in all science, mathematics and engineering (SME) majors" (2000:50). Although university deans and literature intended to recruit students do not necessarily admit its existence, the "weed-out system" has "a semi-legitimate, legendary status and is part of what gives SME majors their image of hardness" (2000:50).

It is pointless to protest against the *principle* of a weed-out system since weed-out systems serve a positive function. For example, the grading system identifies for prospective employers a set of job candidates who are judged the best the school can offer in terms of knowledge, skill, ability, and desire to succeed.

On the other hand, in practice a weed-out system can too easily serve less noble ends since it can be distorted to "insure reproduction of the profession occurs in a way that selects for people with similar social, cultural, and economic characteristics to those already in the profession" (2000:52).

It should now be clear women experience the university differently than men. However, it is necessary to ask a blunt question, "So what?" Does it really matter if women are happy in graduate school?

The answer is that women's happiness matters if their unhappiness is a result of isolation from colleagues and learning opportunities, factors that influence their productivity and commitment to science. As Fox (1999:220) points out:

> In both B.A. and Ph.D.-granting departments, women report significantly less interaction with and recognition from faculty in their departments. In M.A. and Ph.D. departments, the women give significantly lower rankings to the resources available to them and, in Ph.D. programs; they report significantly higher undergraduate teaching loads. The factors of interaction, reported resources, and teaching loads correlate with productivity levels.

Since productivity levels (however measured) ultimately influence career success as indicated by promotions, salary increases, and honors, women's progress toward equal opportunity does hinge on the treatment they get in graduate school as well as in their professional careers. Until women's happiness becomes a priority, in other words, optimistic predictions that they will achieve parity seem hollow since it is not necessarily inevitable women will achieve parity in American science. True, women are achieving entry into graduate school, they are getting doctorates and, yes, some of them are becoming top executives of research universities. Currently, for example, the presidents of Princeton (Shirley Tilghman), University of Pennsylvania (Amy Guttman), Massachusetts Institute of Technology (Susan Hockfield), and Rensselaer Polytechnic Institute (Shirley Jackson) are women and there are others as well.

However, currently in the United States, there certainly is not parity. In many fields in the sciences and engineering, women are less than five percent of the full professors. Across the world, with a few notable exceptions, the ratio of men to women in the STEM fields is four to one or greater (see Table 4). In the United States there is no STEM field where women senior faculty constitute much above ten percent. In short, there is tokenism. It is possible for women to rise to high status in science and engineering but it is *improbable* compared to their male competition. And since the United States needs many more scientists and engineers than it is now producing both the moral imperative and the economic and security needs of the nation combine to make this the right time for universities to become more accommodating to the legitimate needs of their women students. Simply because a few women overcome the obstacles to becoming scientists and engineers is not grounds for complacency. The current system drives out talented people our nation needs if it is to remain preeminent in science and engineering. It damages the lives of many people needlessly and deprives the country of the full benefit of their talent.

However, the problem is not the weed-out system. Some sort of "sift and exclude discipline" as Harrison White (1992) might term it, is essential. The problem is the current version of the weed-out system is manifestly hostile to women. This *perversion* of the "weed-out" system needs to be fixed. Only then will the university be equipped to meet the needs of a society that must tap all its population sectors for scientific and engineering talent.

Before we prescribe what to do, that is, how to make the university more accommodating to women and minorities, we need to have a good understanding about how big the problem is.

Other components of the American population are under represented among senior faculty at major American research universities

Women are not the only component of the population under represented in the ranks of American scientists, especially at the senior level. For example, African Americans and Hispanics regardless of gender are also under represented in the ranks of the nation's scientists and engineers (See Appendix Tables). Gordon (1990) reported more than a quarter (26.3 percent) of African American men receiving bachelor's degrees and almost the same percentage of African American women (23.8 percent) receiving bachelor's degrees in 1986 majored in science, engineering, or mathematics. However, as shown above, few of them moved on to receive doctorates.

As Table 1 shows, in contrast to women, neither African Americans nor Hispanic Americans are significantly represented among persons receiving doctorates in engineering in recent years. In 1994, for example, 2,020 whites of both genders received doctoral degrees in engineering whereas only 54 African Americans and 66 Hispanic Americans received them. In 2001, 1,746 whites received engineering doctorates 92 African Americans and 91 Hispanic Americans received these degrees.

African American and Hispanic women have not sought doctorates in engineering as much as men of those two ethnic categories. In 1994, 305 white women, 119 Asian/Pacific women, 13 African American women and six Hispanic women were awarded doctorates in engineering. During the ensuing seven years, the percentage distribution of doctorates awarded women by ethnic background hardly budged. Black women received from 3 percent to 5 percent of all the doctorates women received in engineering. Hispanic women received from 1 percent to 5 percent of those doctorates. The most plausible conclusion of these data showing as many as a quarter of African Americans major in STEM fields while few go on to receive the doctorate is American institutions of higher learning are not doing an effective job of recruiting and retaining minority students in the sciences and engineering, especially at the graduate level.[5]

Everyone agrees in principle it is essential that our universities begin recruiting large numbers of S and E faculty from the African American and Hispanic communities. However, in the short run, this may not be possible since there are so few to be found. The immediate need is to find out what accounts for limited recruitment and selection of African American and Hispanic men and women for careers in these fields and design appropriate strategies to rectify this problem. It is not only a moral imperative; it is an urgent national task since we need to tap under represented minorities to provide S & E personnel in the future.

Recruiting more under represented minorities or women will necessitate changes in the way universities do business. The universities seem hardly prepared to handle the low numbers of women seeking advanced degrees now. Beyond providing more women's housing, it does not appear universities have given much thought to the ways in which they must revise their approach to education to address the specific needs of women students as aspiring scientists and engineers.

This will have to change. Comfortable myths women students are "honorary men" or tender things unsuited by nature to professional work will no longer suffice. University faculties have to grapple with the needs of this group of students whose brainpower the nation needs. It may be challenging. New courses may be needed. New ways of teaching may be required. Professors may need to take sensitivity courses.

These kinds of changes, not popular in the university especially with the senior faculty, must be seriously considered and implemented when appropriate. We cannot afford to waste a valuable resource by driving women out of science and engineering partly because universities want to continue doing what they have always done. The path followed by university leaders has led the United States to depend on foreign science and engineering talent to compensate for the shortfall in the ranks of homegrown scientific and engineering personnel. Since many of the countries from which these scientists have come, e.g. Brazil, India, Taiwan, are recent competitors in high technology, we can no longer count on them in the future to supply our scientists and engineers. The universities of the United States, white male bastions whose cultures are largely unchanged since universities began to appear in United States during the second half of the nineteenth and early twentieth centuries, must adapt to meet the challenges of the coming decades.

Reform graduate education to increase the number of women scientists with doctorates

Although the details Etzkowitz et al. provide are important additions to our understanding of women students' problems in acquiring the doctoral degree necessary to embark on a career in science, social scientists have been aware for many years of the general difficulties women experience in the process of training to become scientists. Citing literature now over two decades old, J. Cole and Singer (1992, originally 1991) wrote perceptively fifteen years ago about the "formidable early cultural and structural barriers that women face and must hurdle before reaching the starting line" in science. Indeed, they asserted that their theorizing went beyond prior investigators who simply described the gender productivity differences of scientists resulting from differences in men's and women's experiences. Cole and Singer offered a mechanism explaining how these differences in socialization would translate into productivity differences: women in their professional careers suffer greater losses of productivity from negative reactions to their research contributions than men who had less stressful college and graduate school training experiences (1992:285-286).

Perhaps it was not their intention, but by treating women's socialization problems as an exogenous variable in their famous model—i.e., a given—Cole and Singer seemed to imply little could be done about the isolation, and attendant stress, women students felt in graduate school. Women do not inevitably experience these feelings of isolation in graduate school both from other male graduate students and, even sometimes, from each other. When science and engineering faculties want to take effective action, they can make women students feel welcome as equals and can ameliorate if not eliminate their feelings of isolation, self doubt, self blame, and worthlessness. Furthermore, all the evidence from intelligence tests, school grades, and the work of pioneering women scientists demonstrate women can do scientific work the equal of men if given a level playing field.

One of the great strengths of the NAS report is the considerable coverage it gives to innovative programs to help women students overcome their feelings of isolation, self blame, feelings of intellectual inferiority, low self esteem and the like. The NAS identified counseling programs, including peer group programs, and speaker programs that bring in successful women as role models for the women students among other steps universities have taken. One area the NAS admitted it had little information about was how universities might usefully offer sensitivity training both to foreign students about Americans, including women, and about fellow foreign students from other nations. Likewise, American students could learn how to be more sensitive to feelings of the foreign students and junior faculty who might feel just as isolated and who experience some of the same self blame, loss of self confidence and the like as the women students.

An area not covered in the programs reviewed in the NAS survey is the training of university graduate school professors as educators. At many research oriented graduate schools, good teaching is not important to the professors. Even research productivity *per se* is not an overwhelmingly important criterion for advancement. Increasingly, a professor's success in winning grants and doing research yielding patentable products and findings overshadows everything else in determining academic rewards, especially at schools such as Columbia University and other research centers heavily dependent on outside funding.

Oftentimes, good teaching is devalued at the graduate level. Whereas teachers from elementary through high school are trained specialists in teaching, and college teachers at the more selective undergraduate level institutions often are encouraged to be fine teachers (as well as scholars if possible), this is not true at the graduate school level. Students accepted into the program of a graduate school are essentially assumed to be able to learn on their own without any real effort by the faculty to help them. I vividly recall an eminent Columbia University sociologist telling me when I went to see him shortly after being accepted into the graduate program that the professors take little interest in the students in the first two years of their graduate education. Graduate school departments and many large public under graduate colleges often entrust lectures and laboratory teaching to foreign born students and junior faculty.

These individuals, even with the best motivation, have enormous burdens on their time, which limits the attention they can give to their students. Furthermore, some are handicapped by limited English speaking skills, and foreign accents difficult to decipher in a lecture hall environment with dozens of students and less than ideal acoustical properties. Relying on lecturers without good lecturing skills is one *de facto* device large undergraduate schools without rigorous admissions criteria employ to quickly weed out less promising students so they can focus whatever resources they have on the best few.

Summary

Science and engineering undergraduate education in the United States has relied on a weed-out system to identify the most promising students on whom to lavish whatever resources the college has to offer. However, as this chapter emphasized, because of the perverse way the weed-out system operates, women begin leaving science as undergraduates in proportionally greater numbers than men.

Women who survive the undergraduate weed-out system encounter a similar weed-out system in even harsher form in graduate school. For men, the informal social system around a professor generally buffers the worst effects of the weed-out system to a greater degree than for females, who are more isolated and more subject to continued sexual harassment, ridicule, and hostility. Men can often build the contacts and close relationships with others that will serve them well in their post doctoral careers whereas women find few women role models and others with whom they can bond. Furthermore, women students' isolation, when it has not driven them out of science and engineering altogether, will often leave emotional scars and, equally important, lack of a ready pool of potential collaborators with whom to work in their chosen field. That, too, is a serious problem since nowadays, science and engineering, are areas where large numbers of people work collaboratively on projects, and the lone investigator is a quaint anachronism.

Just as the undergraduate weed-out system is worse in large public universities than in small private ones, especially the select schools, the graduate weed-out system does not operate the same way across all disciplines. In psychology, for example, where women have made great strides in reaching the top echelons, the weed-out system does not seem to drive out women disproportionately and as a result large numbers of women armed with doctorates embark on scientific careers. However, in physics, and especially in engineering, the weed-out system as experienced by women is especially harsh and few women reach the starting line from which they will then embark on their careers. In the next chapter I will discuss the obstacles women face in achieving full equality once they reach the starting line, that is, have their doctorates and are ready to begin their careers as practicing scientists.

Endnotes for chapter one

1. In early 2006, Dr. Summers resigned as President of Harvard University calling the school "unmanageable Summers' abrupt departure should not be taken as compelling evidence the University faculty rejected his recanted views about women's lack of aptitude for first rate science. Summers clashed with University faculty over a number of issues, including his pressuring a popular dean to leave.

2. According to Molinari, Betti, Bonfiglio and Mignani (2002), "The percentage of women among students in physics [in Italian university undergraduate programs] has grown from 20.8% in 1960 to 36.4% in 1999." Yet, they also point out, when one looks at the gender composition of the faculty of university physics departments,

> As we proceed along the careers [path], the percentage of women decreases very rapidly. In the university system, the portion of women among those holding permanent positions in physics is 15.3%. [However] among the three tenured levels (approximately corresponding to lecturer, associate professor, full professor), with increasing level, the percentage of women drops from 25.6 [percent] to 15.0 [percent] to 4.9 [percent] of the total.

They further point out "the situation at the highest level is even worse in electronic engineering" where "out of 108 full professors, only two are women."

3. European researchers presented data on the gender gap in the STEM disciplines in November 2005 at a conference sponsored by the Organization for Economic Cooperation and Development. Kamma Langberg, a Danish researcher, compared nine countries in Europe as of 2002 on the "percentage of women among full professors at higher education institutions." She found that Finland, with 20 percent of full professors being women, led all other countries in the proportion of full professors who were women. Austria, for which the most recent data were from 1998, had the lowest proportion of women who were full professors, about six percent. Germany with perhaps seven percent was the next lowest (see Appendix Table A1). Denmark, with just 10 percent of its full professors being women was fourth lowest. She also found that the percentage of professors at Danish universities who were women had risen from just over three percent in 1976 to just over ten percent in 2003.

4. Gary S. Becker, a Nobelist in economics, wrote an essay that appeared in *The Wall Street Journal* on November 30, 2005 reflecting the thinking of the research universities about the merits of highly skilled foreign immigrants. He maintained, "The right approach would be to greatly increase the number of entry permits to highly skilled professionals." The aim would to make all such entry permits permanent. "Skilled immigrants such as engineers and scientists are in fields not attracting many Americans, and they work in IT industries, such as computers and biotech, which have become the backbone of the economy." Becker's proposals need to be examined on their merits. However, I see his statement that engineering and science do not attract many Americans as questionable. The universities, I contend, create a chilly climate for promising women and under represented minorities who might otherwise be happy to apply. There are more than enough of them to fill spots white male students are not seeking in numbers the

research university professors would like. I doubt there is any need over the long term to attract foreign students. Until the research universities make a serious effort to accommodate women and under represented minorities, some short-term stopgap measures to bring in more foreign talent may be unavoidable.

5. Universities are making some efforts to recruit minorities as students with help from the federal government and private foundation sponsored scholarships. It is not certain if programs to recruit under represented minorities will also benefit women students as well. Hispanic societies, for example, have traditionally been highly patriarchal and the possibility exists that increasing numbers of Hispanic men students may exacerbate the problems of women students unless other kinds of improvements occur to ameliorate the hostility women feel in the graduate school environment.

Chapter Two

The mismeasure of scientific productivity

Prefatory remarks

Are women and men equally productive as scientists? The answer to this question, and, indeed, to many others about scientific productivity, is an urgent policy issue. In the last chapter we saw how because of demographic changes—fewer scientists coming to the United States, fewer men majoring in science and engineering than in the past, growing numbers of minority people in the American population—American universities will probably graduate fewer scientists and engineers with doctoral degrees in the future than they do currently unless they manage to recruit many more students than they are currently.

If American universities are going to substantially increase the numbers of scientists and engineers they produce, especially at the doctoral level, they will probably have to recruit from under represented segments of the American population such as women and certain minority groups (African Americans and Hispanic Americans in particular). However, despite some rhetorical commitment to equality of opportunity for all, universities are not rushing to add women and minorities in great numbers.

Indeed, the "question" of whether women are as good scientists as men still gets raised in serious forums. For instance, in early 2005, a flurry of articles in the mass media drew widespread public attention to a speech Dr. Lawrence Summers, the then President of Harvard University, gave to a closed session of a conference on women and science. As reported by Amanda Ripley in *Time* Magazine, President Summers told conference attendees the two main reasons why so few women could be found among the top tier of scientists were (1) women are just not so interested as men in making the sacrifices required by high-powered jobs and furthermore (2) men may have more 'intrinsic aptitude' for high-level science (*Time,* March 7, 2005, p 51).

Summers' comments leaked out and ignited a firestorm of protest both at Harvard and in the editorial pages of numerous popular and elite journals of opinion. Was this merely a tempest in the proverbial teapot? Does it really matter if Harvard's president thinks women are less accomplished scientists than men? (It is important to add here that before his resignation from the presidency Dr. Summers had recanted his views and taken steps to recruit more women for senior positions at Harvard [*The Boston Globe* on February 4, 2005, p 4]).

It would be nice to be able to dismiss Summers' comments as the quaint views of an ivory tower intellectual. Unfortunately, Summers was giving voice to views quite widely shared among the scientific and engineering elite in the United States and many other nations. We are not likely to see those views expressed so baldly in public in the future. However, it is this lack of confidence in women's scientific and engineering prowess (as well as in that of Hispanic an African American people) that motivates the general alarm university leaders express over the difficulties they are having in recruiting foreign S & E talent. Those difficulties are not just the result of less available foreign talent. They are compounded now as a result of heightened security measures taken by the United States in the aftermath of the terrorist attacks of September 11, 2001. One of the latest scholars to articulate that concern is Nobelist Gary S. Becker, an economist at the University of Chicago.

In an essay he published in *The Wall Street Journal* on November 30, 2005 Becker argues we should scrap the current program that provides temporary visas up to five years in favor of unlimited permanent visas for highly skilled professionals. If this occurred, Becker concedes, it would exacerbate the brain drain from other countries unable to pay as highly as American companies, universities, and government agencies can. What he did not say is whether he has in mind only doctoral level skilled level professionals or immigrants with lesser credentials. Currently, the country probably has adequate levels of bachelor's level and master's level specialists in many STEM disciplines but is short of doctoral level specialists in certain critical areas such as bioengineering and information technology.

Mass immigration of even people in critical specialties presents problems Becker has not addressed. For example, what happens when mass layoffs occur in a field? Engineers sometimes have lost their jobs *en masse* as when a large employer such as Lockheed-Martin suffers loss of a huge contract. Social dislocation occurs for many of these people and their families just as it does for people of lesser skills. Would the foreign engineers head back to their countries in these situations and take with them valuable American industrial and military secrets? Who would be responsible for their pensions should the company they work for go bankrupt?

I do not intend to dismiss Becker's plea for some sanity in our immigration policy by raising these questions. Becker's argument should be examined closely on its merits. It does not really matter if it turns out to have been inspired less by a desire to improve America's competitive position in high technology industry than by the reluctance of leading research universities to accommodate women and minority persons in graduate training programs for scientists and engineers.

However, even if we conclude Becker is correct that we must step up recruiting foreign S & E specialists, it is only a *short term* palliative. Ultimately, for the country to have a reliable supply of S & E specialists including top tier talent, it must turn to the native born population. Inevitably, that will force a change in a lopsided ratio of perhaps one female full professor to twelve male full professors in the STEM disciplines. (In physics and engineering, it may be one women full professor to twenty-four men full professors or more)!

Solving the problem of too few scientists and engineers by retooling universities to become more accommodating to women and under represented minorities is the step the present essay advocates. It is a far more ambitious agenda than Becker's. Implicitly, Becker wants this country to accept the reality our universities are incapable of producing the numbers and types of trained personnel our economy needs and we must supplement our output of such personnel by offering incentives to people in other countries to come here.

The agenda proposed here is also far more ambitious than the steps taken so far by many universities. With encouragement from the United States Government many programs now exist at the university level to recruit women for graduate training in the sciences.

The underlying assumption behind these recruitment programs is this will lead to a growth in women in the STEM fields. Perhaps it will result in some increase in junior level faculty and scientists for industry and government positions. However, this essay maintains unless these recruitment efforts are accompanied by fundamental changes in how we evaluate scientific productivity, we will not see any increase in women in top positions in science and engineering. And until women have much more than a token presence in the ranks of science and engineering they will not seek careers in any numbers in those fields. The same is true for under represented minorities.

Becker's policy recommendations and those offered in this essay reflect fundamentally different views about the wisdom, and perhaps, the practicality, of reforming the university system that produces much of America's trained scientific and engineering workforce.

Which approach is correct? If Becker is correct then clearly the American university system cannot change to become more accommodating of women scientists and engineers. Moreover, Becker is counting on the rest of the world to produce a surplus of S & E personnel over the next several decades that might be interested in opportunities in the United States.

His optimistic view contrasts with that of Betty Vetter (1990) who pointed out "13,300 scientists and engineers immigrated to the U.S. in 1970, accounting for 3.6 percent of all immigrants admitted that year . . . in 1988, their numbers had dropped to 10,900, or 1.7 percent of immigrants." Vetter's point reminds us it is not just the attractions of American life, e.g. good income, civil liberties, and good educational opportunities that are pertinent to decisions of foreigners to immigrate. When foreign engineers and scientists consider where to seek employment, they also consider conditions in their own countries and opportunities in other nations.

Universities, government, and industry will need to develop policies and practices that insure the United States produces the trained native S & E workforce rather than counting on immigrants for its needs. The federal government and many state governments recognize this need and have waved the possibility of substantial scholarship assistance in front of the education establishment. This significant carrot is cause for some optimism that in response to constant and well organized pressure over decades, educational bureaucracies can and will change, albeit slowly and unevenly, to accommodate women and under represented minorities both as students and faculty.

The present chapter is organized into two sections. The first section reviews the productivity of scientists and engineers based on the publications count –the current method of evaluating scientific productivity. The main question taken up in this section is: Will the "productivity gap" between men and women disappear? Some renowned scholars—J. Cole and B. Singer (1992), and Yue Xie and K. Shauman (2003) most notably—predict it will eventually. Based on the evidence they and others have presented, this paper reaches a quite different conclusion, that is, their predictions are more wishful thinking than based on the evidence.

The second section raises the question of whether in fact the publications measure is a valid metric of productivity. Implicitly, researchers on scientific productivity, including J. Cole, Yu Xie and Fisher (2005), have asserted it is valid. This chapter questions its validity and urges new measures.

Scientific productivity and selection for the senior positions in science and engineering at major research universities

Who gets selected for the senior positions in science and engineering at the major research universities? Occupants of these prestigious positions come from the pool of doctorates in the appropriate disciplines and, overwhelmingly, that pool consists of white males (though Asians and women are making some inroads, especially in biology, psychology and a few other sciences).

Basically, scientific production is the key to who is selected into the top ranks of science and engineering. As J. Cole and Singer point out (1992:278),

> Science is a highly stratified institution. A small proportion of scientists hold the lion's share of powerful and prestigious positions as well as honorific awards, and this inequality in rewards is paralleled by equally skewed rates of scientific productivity.

The "skewed rates of scientific productivity" to which J. Cole and Singer refer are the number of published books and articles. And they observe a small number of scientists publish a great number while a majority of scientists publish just a few articles apiece. This pattern, J. Cole and Singer note, is as true of women scientists as it is of men.

Yet many more men than women are in this elite group. The reasons for this lopsided ratio of highly productive men to highly productive women are not fully understood. Part of the explanation, as Xie and Shauman show, is a resource gap. Men get more resources—they have more collaborators, they are more likely to have private funding with its less time consuming accountability and application requirements than the government funding on which women are more highly dependent; they are less likely to have heavy teaching loads that reduce time available for scholarly research.

The resource gap is not the only challenge women face. The publications count, the commonly used yardstick for productivity in science, itself is a defective measure undercounting women's productivity, even if the undercount is not deliberate. J. Cole and Singer themselves equate scientific productivity with "the numeric count of published articles and books

Despite numeric data on doctorates in science and engineering suggesting perhaps scientists forge the ties to help them become productive scholars while still in graduate school, and women are disadvantaged in this respect, little attention in the sociological literature on productivity has been given to the issue of social factors in the formative years of a scientist or more generally to the processes of recruitment and selection into the entry level positions in these disciplines that may influence later productivity. With the notable exception of Xie and Shauman's study *Women in Science* (2003), there are few empirical works on this issue and little theorizing to guide scholarship.

Sociological interest in scientific productivity has centered on the "gender gap" in publications, i.e., the differential between men and women scientists launched on their research careers after receiving their doctorates. By 1984, for example, J. Cole and Zuckerman had counted "more than 50 studies" of scientific productivity differences between men and women scientists (1984:218; also see Xie and Shauman 2003:176).

Many more studies have been done since the mid 1980s on the gender gap in productivity, making it one of the most studied phenomena in the social sciences. While nothing can compensate for the paucity of scholarly attention to processes of recruitment into science and engineering (since without good information on recruitment into science we cannot hope to understand the reasons for the lopsided ratio of men to women faculty at the senior level of major research universities), we should welcome good research on the "gender gap" in science for what it tells us about the experience of the relative handful of women who have pursued academic careers in S and E disciplines. Unfortunately, despite the large number of studies of the gender gap, there is little empirical research on the *causes* of the gender gap in publications. Much of the material available on causes is anecdotal and theoretical.

Despite the poor state of knowledge (or perhaps because of it?) scholars in this specialty have sharply divided views they defend passionately. For example, scientists continue to debate all the following: (1) what is the definition of scientific productivity? (2) Is there a gender gap in productivity? (3) Will the gender gap in productivity disappear (or has it disappeared already at least in some disciplines)?

It seems useful to address the last question first, if only because if the productivity gap is going to disappear anyway why bother to debate either its causes or its consequences? We merely need to sit back and let the problem disappear while we address other matters where active intervention is required.

Is the productivity gap closing?

Scholars taking this position rely on two pieces of evidence for their prediction: (1) women scientists have closed the publications gap in some fields such as sociology (Valian 1998; also Stack 1994) and economics (Kolpin and Singell 1996); and (2) over the past thirty years, women's productivity, measured as a proportion of men's publication rate, had climbed to 82 percent in 1993 from 58 percent in 1969 (Xie and Shauman 2003).

In regard to the first point, there is no reason to debate women are as productive in sociology and economics as men. However, it is a dangerous logic fallacy to reason from these empirical findings in two disciplines to many others when there are no grounds to believe the two disciplines are at all like the others. Before we can be confident women's progress in two fields are harbingers of the future we must first ask: Are the processes affecting productivity identical for all fields? Do we know what they are?

At least one authority, Mary Frank Fox (1995, citing Zuckerman, 1987) observed, "the factors governing variations in women's employment by field, particularly the higher proportion of women in life compared with physical sciences, are not well understood."

Simply because we have evidence women produce as much as men in a couple of fields it is absurd to assume they will catch up elsewhere. Much more needs to be known about the particular disciplines and women's position in them before we can speak confidently about the productivity gap disappearing in them.

In regard to the second point, that is, women's productivity appears to be catching up to men's, Xie and Shauman, to their credit, investigated fields where women have had difficulty penetrating the top ranks: engineering, physics, biological sciences and mathematical sciences (2003:179).

However, two facts undermined Xie and Shauman's argument the gender gap is disappearing. First, as their Table 9.1 shows, the mean cumulative publications count shows women produced far fewer publications than men: 22.9 versus 29.65 for men (2003:182). Second, although the "total publications in last two years" is close for men and women (4.47 for women versus 5.47 for men), it still favors men. Furthermore, since the short-term measure is not validated against some other independent indicator of women's greater productivity in the past two years, it is open to challenge on methodological grounds as an indicator of productivity. The basic problem is a measure such as two-year productivity is subject to irrelevant response bias. In short, Xie and Shauman do not make a compelling case the productivity gap is likely to close anytime soon in fields other than sociology and economics. They do make a persuasive case, as stated above; a resource differential is behind much of the publications gap though perhaps not all of it.

If the productivity gap is not disappearing, at least as measured by the publications count, what should we do? Elizabeth Ivey, the president of the Association of Women in Science has pushed the idea of "publish earlier and more often" as a trick of the trade women need to learn. "Women researchers don't tend to publish their results until they're very near the end of their project, whereas male researchers will publish intermediate results all along the way, so they build maybe three articles on a project where women tend to have only one" according to Ivey (2005).

Ivey's intent to mentor women as a means to improve their scientific productivity and success is laudable. However, urging women to publish more papers may be of little practical value because as Claudine Hermann, the only woman full professor in the physics department at France's prestigious Ecole Polytechnique explains, "Wenneras and Wold . . . demonstrated . . . a woman had to publish 2.6 times more than a man to be awarded a grant by the Swedish Medical Research Council." (Also see Wenneras and Wold, 1997). Similar biases were evidenced in Denmark (Hermann 2002:76).

Furthermore, as a blue ribbon panel of women leaders in education and scientific research remind us, "When evaluators rated writing skills, resumes, journal articles, and career paths, they gave lower ratings on average if they were told . . . the subject of evaluation was a woman" (Handelsman et al. 2005:1191).

Ivey's suggestion shows her recognition of the inherent lack of validity of the publications count as a measure of productivity. However, her proposal does not recognize men exhibit gender bias when assessing the worth of fellow researchers' work. Specifically, as Barres tartly observed (Begley 2006), men do not count contributions by men and women the same way when they know the gender of the scientist whose work they are assessing. That is one major reason why we need to develop new tools for assessing scientific contributions rather than continue to rely on the current methods, especially the publications count, when evaluating scientists for tenure, promotion, and academic honors.

Two other approaches might be considered to redressing the lopsided ratio of men to women senior faculty in the STEM disciplines, especially in the leading research universities. Possibly, the problem can be alleviated either by changing the conditions under which women work or, alternatively, by changing the indicator of productivity. In the next section the first option will be examined. I will save for the next chapter a discussion of the second option.

Can the working conditions of women be improved as a means of improving women's publication count? One group of scholars, whom I label the "resource differential" camp would probably answer in the affirmative. Another group, which emphasizes women's socialization as the primary factor, is not so certain. It is important to be familiar with each camp's position since the debate is hardly settled.

Mary Frank Fox (1995) and Yu Xie and Kimberlee Shauman (1998:2003) are noted proponents of the view resource differentials are the main reason for gender differences in publication rates. The resource differentials are not necessarily gross differences such as the sponsor's willingness to pay men researcher higher salaries than women or to give men more money than women for supplies to do similar work. Subtler forms of resource disparity can arise such as differential membership in the invisible colleges where research problems are often first identified. As Mary Frank Fox (1992:195) remarks, (quoting J. Cole 1981), "although women have moved into science they are not *of* the community of science." Women more often than men remain "outside the heated discussions, inner cadres, and social networks in which scientific ideas are aired, exchanged, and evaluated."

The essence of this camp's argument, as summarized in Xie and Shauman (2003:192), is that men have "the personal characteristics, structural positions, and facilitating resources . . . conducive to publication." Xie and Shauman admit they do not know why men and women differ in these dimensions, implicitly conceding the resource differentials are not necessarily the result of gross forms of discrimination.

J. Cole and Singer are the most notable proponents of the view that socialization is key to understanding gender differences in productivity. Their basic thesis is contained in their now classic paper "A Theory of Limited Differences: Explaining the Productivity Puzzle in Science" (published in 1991 and reprinted 1992). The gist of their position is contained in the following observation,

> Socialization processes lead young women to be less confident about their scientific ability, less assertive in advancing their ideas and opinions, less apt to pursue their goals aggressively, while simultaneously being more ambivalent than men about their work and family roles. In due course, women and men come to the starting line for scientific careers carrying baggage of substantially different weights. Differences in scientific production follow naturally from these differences in background and current attitudes and traits (1992: 285).

A few scholars, notably Virginia Valian (1998), have taken the position that both camps have some good points to make. My own position is somewhat similar to Valian's in one respect, that is, I also accept as correct some points made by both camps. Unlike Valian I emphasize the problem choice process's influence on productivity, e.g. the influence of gender differences in a research problem's feasibility and potential intellectual payoff prior to commencing the research (see Fisher 2005).

Productivity differences: Are they the result of resource differentials or of socialization differences?

Much of the discrepancy in productivity between men and women consists in the much higher number of publications of a relatively small number of men scientists—the kind who attains the senior faculty ranks of major research universities (Valian 1998; also see J. Cole and Zuckerman 1984). Valian says

> The average difference in productivity between men and women is partly, but not completely, due to the fact . . . there are more men than women with extremely high publication rates. . . . In most fields, 10 to 15 percent of the scholars account for 50 percent or more of the publications (1998: 262-263).

Further depressing the overall publication rate of women, as Valian remarks, is the fact "women are . . . more highly represented among those who do not publish at all" (Valian 1998:263; also Long 1992).

The difference in numbers of men and women with extremely high publication rates is an interesting phenomenon. First, it bears directly on the question of why so few women proportionally in senior faculty positions at the major research universities. It is these institutions of higher learning that attract the largest proportion of the most productive of scientists. Second, the question is intriguing by itself since it may help illuminate why women generally publish less than men regardless of discipline.

One example of a resource differential helping to account for men's greater rate of publication is men have more collaborators in their work. Cameron (1978); also see (Fox 1995) and more recently Bozeman (2004) found this difference. Fox (1995:221) suggests, "Women may have more difficulty finding and establishing collaborators and they may have fewer collaborators available to them."

Data on the state of women in physics abroad, summarized in Hartline and Michelman-Ribeiro, Eds. (2005), leave no doubt resource differentials are a serious problem in many societies.

Kodate and Torikai (2005) reported in Japan women face "considerable disparity . . . in the allocation of research resources, including funding." In a survey conducted in 2003, conducted by a "Committee for Promoting Equal Participation of Men and Women in Science and Engineering" of which Torikai was a member, she stated the Committee found, "for example, 11% of men but only 3% of women have annual laboratory budgets of over 20 million yen. . . . Many women reported unfair treatment after taking child-care leave, and remain under funded throughout their careers."

In Belarus, according to Svirina et al., "Women physicists in Belarus meet the same problems in adaptation and self-realization as their foreign colleagues (less equipment, laboratory space, and salary, and more administrative and psychological pressure)."

Xie and Shauman, however, marshal some of the most impressive evidence of chronic resource differentials in several areas based on survey data for the United States. Especially noteworthy are the data they report in Table 9.2 (2003:188) and Appendix Tables D9.a-D9.d. For example, according to Xie and Shauman's Appendix Tables D9.a-D9.d, in 1969, adapted here (see Table 6 below) 37 percent of women had teaching loads of at least 11 hours—the heaviest category—versus 18 percent of men. In 1973—where the data were reported somewhat differently—I estimate it was just fewer than 30 percent of women versus 17 percent of men. In 1988 it was 29 percent of both men and women and in 1993 it was 31 percent of women and 23 percent of men.

Table 6: Selected Indicators of Resource Differences Between Men and Women Scientists

Resource Differentials Between Men and Women		
a. Teaching Loads of Eleven Hours and Above		
Year	**Men**	**Women**
1969[1]	18%	37%
1973[2]	17%[5]	30%[5]
1988[3]	29%	29%
1993[4]	23%	31%
b. Research Assistants		
Year	**Men**	**Women**
1969[1]	53%	22%
1973[2]	50%	30%
1988[3]	77%	71%
1993[4]	77%	73%
c. Full Professorships		
Year	**Men**	**Women**
1969[1]	39%	26%
1973[2]	44%	26%
1988[3]	52%	26%
1993[4]	51%	30%
d. Supported by Industrial / Private Foundation Grants		
Year	**Men**	**Women**
1969[1]	30%	10%
1973[2]	20%	10%
1988[3]	23%	23%
1993[4]	33%	19%
[1] Source: Xie, Yu and Kimberlee Shauman (2003) Table 9.a. (p. 268).		
[2] Source: Table 9.b (op. cit.) p. 270		
[3] Source: Table 9.c (op. cit.) p. 271		
[4] Source: Table 9.d (op. cit.) p. 274		
[5] Interpolation of data reported in Table 9.b (op. cit.) p. 270		

Although the gap has certainly diminished over the 24-year observation period women still were carrying more of the teaching burden at the end of Xie and Shauman's study period.

In the case of "research assistants" again, the differences narrowed a lot over the 24-year observation period but did not close. In 1969, 22 percent of the women had research assistants, versus 53 percent of the men. In 1973, 30 percent of the women had research assistants and 50 percent of the men had them. In 1988, 71 percent of the women had them while 77 percent of the men had them and in 1993, 73 percent of the men versus 79 percent of the women had this resource.

Income differentials between men and women in academic sphere probably exist, at least at the full professor level, but data are not easy to find. However, it is possible to look at the relative professional standing of the two genders, which correlates highly with income. This comparison shows the gender gap in the percentage of each gender with full professorships has widened from 1969 to 1993. In 1969, the percentage of women and men who were full professors was 26 percent and 39 percent respectively. In 1973, it was (again) 26 percent and 44 percent respectively. In 1988, it was (still) 26 percent and 52 percent, respectively. Finally, in 1993, it was 30 percent and 51 percent respectively. Over the observation, period women had increased their representation in the full professor ranks by less than five percent, whereas men had increased their representation by about 12 percent, raising the gap from 13 percent in 1969 to 21 percent in 1993.

Sonnert's study (1990) which looked at recipients of "prestigious postdoctoral fellowships awarded between 1955 and 1986" is apposite in this connection. According to Mary Frank Fox (1995:214), Sonnert found "the predicted academic rank of women is one-third lower than that of men" after controlling for "years since doctorate, fields, and fellowship." When he looked at the results by field, he found the "disadvantage for women" was in "physical science, mathematics, and engineering"—three of the fields included in Xie and Shauman's study—but not in biological sciences.

Finally, there is a gap in the proportions of researchers supported by private nonprofit and industrial grants. According to Xie and Shauman, in 1969, 10 percent of women received funding from industrial grantors and private foundations combined. This was half the 20 percent of men similarly funded. The proportions did not change in 1973. In 1988, women caught up with men—23 percent of women received such funding versus 22 percent of men. However in 1993, 19 percent of women received private foundation or industrial grant funding versus 33 percent of men.

It is unfortunate Xie and Shauman could not locate data showing a disparity in level of funding as had been found, for example, by Torikai in Japan. Compared with the data Xie and Shauman reported on percentage of each discipline's full professors who are women versus men or their data on percentage of each gender carrying the highest teaching loads, for instance, Xie and Shauman's data on source of research funds are more difficult to interpret.

However, I maintain the source of funding, private versus government, can be useful for shedding light on the disadvantages women face in securing funds for their research. Private funding is more desirable in some ways than government grants. Researchers dependent on government grants spend a great deal of time on writing proposals and dealing with government accountability requirements. As Stephan remarks (1996:1226), working on grant applications diverts scientists from spending time doing science. The time expenditure is considerable since a "funded chemist in the U.S. can easily spend 300 hours per year" just on proposal writing" (1996:1226).

The greater proportion of women than men dependent on federal and other government funding may be a contributing factor in lower productivity of women researchers. Furthermore, it is important to remember many scientists experience more than one deficit. The same woman scholar dealing with burdensome government paperwork requirements on her grant may also have less desirable committee assignments in her department than men of equal rank. Or she may have not had a research assistant assigned to her by the department.

Using a sophisticated multivariate analysis, Xie and Shauman convincingly demonstrated resource differences make a difference in productivity. Xie and Shauman tested their hypothesis about the effects of resources by looking at looked at productivity in "biological science, engineering, mathematical science, and physical science" (2003:179). They chose those four disciplines to "make our results comparable with those from earlier research."

In their Table 9.2 Xie and Shauman showed the "short term productivity" of women as a percentage of men scientists' increased in the interval 1969 to 1993 from 58 percent to 82 percent. Since gender was invariant, any increase in productivity could not possibly be attributable to this factor.

Xie and Shauman then added several resource measures. Model 2 shows the effects of the new variables most clearly. Controlling for their *combined* effects (equivalent to having treatment groups similar on these variables but differing on gender) resulted in an astonishing change. In 1969, the women's publications rate was 95 percent of men's, in 1973 it was 94 percent and in 1993 it was 93 percent. For unclear reasons (a possible drop in federal grants differentially affecting women?), women's publications rate was 78 percent of men's in 1988.

Although gender, a proxy for unknown variables, had an extremely strong effect as a predictor in Xie and Shauman's study, they reported including the variables of "discipline", "time between BA/BS and PhD", and "years of experience" significantly improved the model's explanatory power.

Then Xie and Shauman added to their analysis several resource variables shown in Table 6 above. Although the ratio of women to men's productivity jumped, the new model was not statistically significant. How do they explain this puzzling result? Xie and Shauman insist the relationship of resources to productivity, discussed in their 1998 paper (Xie, Yu and Kimberlee Shauman 1998), has not declined. In fact, they believe it *strengthened* over time. Xie and Shauman attribute the lack of significance in the model to "the fact . . . these resources become more equally distributed between men and women." In part, this argument seems correct as Table 6 shows. However, their contention also may be *incorrect* regarding such resources as full professorships or private foundation/industrial grant support where the gender gap appears to have widened according to Table 6.

Table 7: Estimated Female – Male Ratio of Publication Count for Three Negative Binomial Models of Research Productivity (Carnegie -1969, Ace-1973, NSFP-1988, NSFP-1993)[1]

Estimated Female – Male Ratio of Publication Count				
Model Description	**1969**	**1973**	**1988**	**1993**
Gender	0.580[2]	0.632[2]	0.695[3]	0.817
Gender + Field + Time between BA/BS and Ph.D. + Years of Experience	0.630[2]	0.663[2]	0.800	0.789[4]
Gender + Field + Time between BA/BS and Ph.D. + Years of Experience + Rank + Teaching Hours + Research Funding + Research Assistants	0.952	0.936	0.775	0.931
[1] Source: Table 9.2 of Xie, Yu and Shauman (2003) p. 188				
[2] p< .001 (two tailed test of the hypothesis that research productivity is the same for men and women)				
[3] p< .01 (two tailed test of the hypothesis that research productivity is the same for men and women)				
[4] p< .05 (two tailed test of the hypothesis that research productivity is the same for men and women)				

The simplest explanation of their findings—and most plausible—is the underlying relationship of resource variables to productivity was *invariant* over the nearly quarter century observation period. That is, if men and women were given equivalent resources they would have performed comparably on a *valid* productivity measure in the observation interval. The inverse is also true; the true relationship of resources to productivity would not be detected if productivity were not measured validly. In fact, if "publications in a given two-year period" (the Xie and Shauman productivity measure) was becoming more unreliable over time—as would be the case if it suffered from irrelevant response bias —Xie and Shauman would have found exactly this kind of "change" in the measured relationship of resources to productivity. In other words, as the indicator of productivity became more unreliable, quite possibly a "change" in the relationship of resources to measured productivity would be seen in the data. Xie and Shauman, who have argued that the gender gap is disappearing, do not share my interpretation of their findings.

Despite the problems in the productivity measure, the resource differences camp, in my opinion, make a persuasive case; resource differentials between men and women scientists throughout their careers play a significant role in explaining any real productivity differences between them. However, this view has been disputed by a number of other scholars and their criticism is important to understand. The most sophisticated criticism comes from J. Cole and Singer (1991; reprinted 1992). Cole and Singer do not deny the resource differential between men and women scholars. Rather, they challenge the argument that resource differentials *cause* lower productivity. Women receive fewer resources for research because they are less productive and they also receive fewer rewards in the form of academic honors and promotions etc. for the same reason. The reason for women's lower productivity, Cole and Singer (1991, reprinted 1992) suggest, is because of limited differences in socialization.

Cole and Singer emphasize the lack of encouragement women receive to enter science and engineering throughout their formative years. For example, women who show mathematical aptitude may not be encouraged to pursue careers in science and engineering while in college. Instead they may be pressured to go into teaching (a fine profession by itself but not a research discipline). When women with aptitude for science and engineering nevertheless head for graduate school, they may find fewer top professors to supervise their doctoral research.

Much anecdotal data support these observations by Cole and Singer. Cornelia Dean, a *New York Times* science editor, recalled her own decision to become a science editor. Dean said she was "scarred for life" when, as a seventh grader at an "experimental state school for brainiacs", she took a mathematics aptitude test. "The results", she remembers, "were posted and everyone found out I had scored several years ahead of the next brightest kid." Her reaction to this situation is sobering. She remembered thinking to herself, "A girl really good in math! What a freak!" and she "resolved then and there on a career in journalism."

J. Cole and Singer go beyond cataloguing the assorted slights, the verbal harassment, and outright hostility to women who show promise in mathematics and science. Their model emphasizes a key to women not being as productive is their reaction to these slights, that is, women scientists may react to these disappointments by becoming discouraged and lowering their expectations. Men faced with these disappointments are less likely to be discouraged and do not lower their expectations as much. This difference in reactions to the slights, harassment and other manifestations of hostility accounts for gender differences in scientific productivity.

As intuitively appealing as J. Cole and Singer's model is, it is not well grounded in data. Surprisingly little systematic research has been published on such issues as what factors cause students to choose majors in college. One recent study (Lackland 2001) found "students' value systems . . . were a significant predictor of major choices." Lackland further noted "endorsement of humanitarian concerns was associated with selection of a helping profession major" (e.g. teaching, nursing) while "failure to endorse humanitarian concerns was associated with selection of a science major" (2001:5).

Lackland astutely observed, "College personnel need to emphasize the full range of values associated with all fields of study as a means to increase diversity in the major pool" (2001:6). She pointed out while teaching attracts people with strong humanitarian concerns, it also could appeal to people with weaker humanitarian concerns and strong utilitarian concerns if student advisors emphasized the ten-month year, the steady pay, and strong job security for science and mathematics teachers. Similarly, guidance counselors can advise students with aptitude in mathematics and engineering, but also with a strong humanitarian bent, as engineers they can work on projects advancing health care and public safety. These disciplines have plenty of room for people with humanitarian values.

Although studies such as Lackland's are beginning to uncover the complex social processes connected to choice of a major and decisions about what post graduate education to pursue, we still lack systematic information on socialization for science careers. This is the main stumbling block to understanding why fewer women and hardly any Hispanic and African Americans of either gender decide to pursue graduate work in engineering and the dispassionate sciences of physics and chemistry.

However, the Dean anecdote points to the possibility it is *not* an accumulation of slights throughout their school years that may be pertinent to career choices, but rather a life-defining or life-altering incident at a *critical moment* in their childhood or formative years. Perhaps it is the winning of a science fair prize that fires the ambition of a youngster to become a chemist or, alternatively an embarrassing, or scary, incident that squelches an ambition to become a mathematician. In Dean's case she retained an abiding interest in science—after all she eventually became a science editor for a leading daily newspaper—but she did not pursue a doctorate in science or engineering because she felt, at a time when acceptance by her peers was a foremost concern, demonstrating her aptitude in mathematics would brand her a "freak." And this was at a school for gifted children!

While their understanding of the dynamics by which young people make their science career decisions may be incomplete or erroneous in some ways, Cole and Singer's hypothesis is appealing because it may point to differences in the way men and women react to verbal and physical aggression. From an early age women are more attuned to speech. They are more likely to try to placate a person who shows hostility to them whereas men are more likely to react with anger and aggression to displays of hostility. To put the gender difference plainly, we might imagine a male college student told by his chemistry professor he will never be a scientist decides to redouble his efforts to prove the professor incorrect whereas a girl told the same thing might seek to become a high school science teacher instead or change fields from the sciences to history.

While Cole and Singer emphasize the importance of scientist socialization, and the resource differential scholars largely ignore this issue in favor of post career resource differentials, close scrutiny of the two positions shows considerable agreement between them. The resource theorists, such as Mary Frank Fox on the one hand and Xie and Shauman on the other, do not deny women are hurt by slights, insults, and other aggression in their formative years. However, they have argued persuasively real differences in resources for research have an *independent* effect on women scientists' productivity beyond any effect from psychological damage women have suffered during their formative years. For their part, Cole and Singer (1992:279) acknowledge "simple discrimination" has played a role in the "gross disparities" between men's and women's productivity but assert this alone is not decisive.

If the differences between the two camps are not that large why is the debate so heated? Part of the reason, of course, is academics naturally take offense at what they regard as unfair attacks on their character, motives, or work.

Indeed, much of the animosity between adherents of the "resource deficiency" and "limited differences" camps has less to do with the intellectual merits of the various positions than with mutual suspicion each side harbors toward the other. Adherents of the resource differential position claim the Cole and Singer position gives short shrift to resource differences between men and women after scientists launch their careers. The "limited differences" adherents counter the "resource differential" camp is too quick to accuse university administrators and male professors of gender bias, when the problem is really a particular scholar's deficiency.

When you look closely, however, the resource differential camp is less concerned with what the Cole and Singer thesis says than it is with how the Cole and Singer thesis could be misused. Basically, the resource differential camp worries the Cole and Singer thesis will be used to imply wrongly women are responsible for their lack of progress in rising to the top in the academic world when often women are caught in a Catch-22. According to resource differential camp adherents, too often academic department chairmen and other top officials treat women scholars as unable to do top quality work. The women do not get resources in a timely way or in quantities they need and the women then do not do as good a study as possible. The department heads then blame the women for the inferior work they did, resulting in delays in getting tenure, promotions, and academic honors. Therefore, the resource theorists suspect the Cole and Singer camp scholars are apologists for academic employers who discriminate against qualified women candidates for promotion, salary increases and academic honors.

The Cole and Singer camp bridles at any suggestion they are apologists for irresponsible administrators who discriminate against women scholars. Adherents of the Cole and Singer camp wonder, however, if the resource differential theory camp is all too eager to see universities and colleges as villains when, in fact, many women researchers' problems in reaching the top of their field may be attributable to problems in their formative years for which their university and college employers are not responsible.

I agree with the Cole and Singer camp we should not presume at the outset university decisions that are unfavorable for particular women in matters of tenure, promotion, salary, or honors reflect some general hostility to women. I also reject the notion the Cole and Singer hypothesis is an attempt to clothe a "blame the victim" ideology in academically respectable garb. I see nothing wrong with the idea women at the start of their careers already have learned some ways to protect their work from critics who suspect women cannot do quality work. Regrettably, some of the defensive measures women have taken, e.g. a greater likelihood than men to study problems renowned researchers have suggested are worthwhile, slow down the problem choice process and delay publication—negatively affecting women scientists' publication rate.

Although I consider myself a supporter of the Cole-Singer model, I support a *modified* version of it. For example, I share the resource differential theorist view resource differentials make an independent contribution to scientific productivity. In my opinion, there is two-way causation in the relationship of productivity and resources. The process as conceptualized here unfolds roughly as follows: women, starting their careers with fewer collaborators (see Bozeman 2004; also Fox 1995:221) and isolated from "the intense discussions, inner cadres, and social networks in which scientific ideas are aired, exchanged and evaluated" (Cole 1981; Fox 1992) begin their careers with several subtle disadvantages in the STEM disciplines that have an adverse impact on their productivity, especially in the form of publications. Men are less likely to have these disadvantages at the outset. Slowly men build a higher publications record. Demonstrated high productivity (i.e., many publications) leads to more resources being awarded; more resources will permit the scientists to be more productive in the future. Lower productivity (i.e., few publications) leads to less resources being awarded; this can limit productivity in the future. Women are more likely to be in this lower productivity net than men.

Number of publications: A defective yardstick

In the academic sphere the two primary measures of research performance are quantity and quality, according to Virginia Valian (1998). While quality of work has always figured into assessments of scientific importance, no agreed upon definition of quality exists. Some writers on productivity, e.g. Valian, assert if "quality", as measured by citations per paper, was considered, women's productivity would be more accurately assessed. However "citations per paper" is not a satisfactory solution since, among other drawbacks, it has problems of interpretation. Is a much cited work better than a neglected paper with important insights? For example, Is Bardeen-Cooper-Schrieffer theory of superconductivity, which influenced most physicists for a long time before being scrapped, better than Gregor Mendel's pioneering work in genetics, which was ignored for decades? When should we decide the issue of how good a paper is in science, based on citations per paper?

Quantity is typically measured as a simple ratio of "numbers of publications" divided by "a unit of time." However, as Valian noted, "the count is not completely straightforward" (1998:261). Valian observed that just as a lengthy article reporting the results of a series of important experiments is a publication, a short article reviewing a book is also a publication. Valian reasonably asks, "Are all articles equal? How many articles equal one book?"

Specialists in the area of scientific productivity are quite aware of the litany of problems of the publications count. Certain customs within particular disciplines cause the count to be inflated. It is common in biological sciences to credit the person who has permitted use of his biological materials as an author. In most fields, every professor with a large grant supporting junior level researchers is a coauthor of many papers done by his/her students for refereed journals.

Men have been the primary beneficiaries of some of these easy ways to enhance one's publications rate. For instance, men benefit from counting coauthored papers more than women because men have more coauthors (Bozeman and Lee 2004). And men probably also have benefited more than women from the practice of counting suppliers of biological materials as coauthors.

Specialists in the area of scientific productivity are quite aware of the litany of problems of the measure they use. However, while bemoaning its limitations, they continue to use it. No real movement is occurring to replace or even supplement the publications measure. Consequently, in science and engineering, an individual scientist's standing among his or her peers hinges, to a considerable extent, on how many publications he or she has authored.

Lopsided ratio of men to women will not change quickly in response to changes in women's productivity relative to men's

Let us suppose for the moment women's scientific productivity improves, as measured by publications. And further, let us suppose the universities make a sincere concerted effort to reward women for this improvement in their productivity. The simple fact is a dramatic change in the ratio of men to women senior faculty will not result in the near term.

Pointing to the "institutional arrangement" for promotions in academe, Jerome Bentley, an economist who has looked at the question of how to diversify the science and engineering faculties of American universities observed, "The whole process takes so long: The typical tenure track is six years, and generally there's a lag between associate professor and full." Moreover, he added, "If you have a workforce that's primarily white male and they're tenured, you can't fire them. You have to wait" (Fodor, 2005:3).

The wait will be long now that mandatory retirement ages no longer exist. Professors today are active well into their seventies and beyond and fewer universities are financially able to expand the number of professorships to accommodate women scientists.

The senior professors, moreover, control the distribution of resources within the department and will generally be a force for continuing to do things as they have been done in the past. It is likely the resource gap, which contributes to the productivity gap and to the lopsided ratio of men to women, may not disappear soon because the men who control the resources will generally prefer to hand on that control to younger men with whom they have enjoyed good professional and perhaps personal relationships.

Chapter conclusions

Without a fundamental overhaul in how we assess scientific productivity, we will never be able to have equality of opportunity for men and women in science. Achieving equality of opportunity is not simply a question of elementary justice—and on those grounds alone, a matter of great importance. It also has been recognized for sixty years or more to be a matter of national importance to the economy and security of the United States. Vannevar Bush (1945) foresaw the day when people would be judged solely on merit and science and engineering would welcome their fair share of able people regardless of gender or ethnic background. However, that has not happened yet.

Can this ratio of women to men full professors ever be increased to about one to one? There is reason for guarded optimism in the *long* term. In some fields such as psychology women have made considerable strides and quite possibly in the future this will show up as a higher ratio of women senior professors to men professors, perhaps above 20 percent in the next two or three decades given current trends in the production of doctorates and the existence of a significant group of senior women scholars already in the field who can act as role models for other young women. In engineering and physics, where the base of senior women is tiny, achieving a large percentage of women senior faculty will take a very long time, perhaps fifty or more years at a minimum, thanks to lack of mandatory retirement ages, pinched university budgets, and the difficulties in attracting women to fields historically regarded as men's work.

Some scholars including Xie and Shauman (2003) and Cole and Singer (1991) point to data showing a narrowing in the publications gap and proclaim this presages an overall improvement in women's status in the coming decades. However, they are vague as to when this will occur and the evidence is weak it will happen in most of the STEM fields in the foreseeable future. The best hope for growth in the numbers of senior women scientists is expansion of the total number of professorships and efforts to fill those slots with qualified women and minorities.

One way to accomplish an increase in the number of women senior faculty is to change the criteria used for assessing productivity. It will be difficult to do. Despite the defects of the quantity measure of productivity, the measure most widely used, it enjoys widespread support for a host of reasons. One is that the publications rate is simple to compute (publications divided by a unit of time). Second, this measure is familiar and people are comfortable with it because it has some intuitive appeal. Third, on the surface it does not appear to discriminate against women (though counting coauthored papers probably helps improve the productivity scores of men more than women). Finally, prominent scholars in both major camps (Cole and Singer 1991; Valian 1998 and Xie and

Shauman 2003) seem to be convinced that *eventually* the publications gap will disappear by itself even if we never really understand what caused it. I do not share this confidence in the eventual disappearance of the problem, but it seems to be an incubus to developing new productivity measures.

Regrettably no one has given serious consideration to replacing the publications count, a measure everyone concedes is flawed, nor to the need to ask a different question(s). Even among women in science the quantity measure is largely accepted as something that will always be used. Elizabeth Ivey, the president of the Association of Women in Science, has pushed for women to publish earlier and more often. "Women researchers don't tend to publish their results until they're very near the end of their project, whereas male researchers will publish intermediate results all along the way." Men's greater savoir-faire in this regard allows them to generate "maybe three articles on a project where women tend to have only one" according to Ivey. Would Ivey make this suggestion if she or other thoughtful and committed activists on behalf of gender equality believed new productivity measures were a real possibility?

The legal climate has changed and universities are under pressure to open doors to women and minorities. So far, however, the universities have been able to defeat the intent of the law by recruiting S and E talent from so-called "third world countries" who meet the legal definition of under represented minorities even as they exacerbate the problems of women in the professions. As remarked earlier, this is because many of these foreign scientific and technical personnel arrive in the United States with attitudes hostile to equality in the work place for women. (see NSF 1991, also Rosser 2004 and Etzkowitz 2001)

However, with the difficulties of recruiting foreign talent growing, universities are going to face the problem of adapting to make a more welcoming environment for women and U.S. born minorities. The universities can do this intelligently by putting into place a long-term strategy that leads to enhancing their numbers of women faculty members by a *transparent* assessment process based on *job relevant* criteria. Or the universities can do this in counter productive ways that breed resentment against less qualified women getting "priority" over more qualified men.

As we have seen the current measure of productivity is harmful in perpetuating a lopsided ratio of men to women in academia and a shortfall in the number of able students who recruit into careers in science and engineering at the doctoral level. (It is disputed whether the country has too few or too many engineers and scientists at bachelor's and master's level). However, we should not forget the country has even more severe problems in respect to recruiting, retaining, and promoting minorities such as African Americans and Hispanic scientists regardless of gender.

Gender neutral criteria should help overcome the problem of too few women from various components of the population of the United States in the top ranks of science but may not address the problems of men from these populations. Yet, in becoming sensitive to a variety of "micro-inequities" as well as more dramatic deficiencies in the recruitment and advancement process academic administrators may take other steps that will bring more men from these under represented groups into the academic world.

Chapter Three

New ways of conceptualizing scientific productivity

The previous chapters have documented the foibles and deficiencies of the STEM fields. That is the easy work. Fixing the problems will be more difficult and time-consuming, and the fixes will often be imperfect.

Other writers on this subject have emphasized the need to improve recruiting of new talent from "under represented minorities" or training women in ways to enhance their publications count. I call for an entirely different approach that makes paramount reforming the evaluation system in science and engineering for practicing disciplinarians.

The future belongs to the society best adapted to meet the needs for technically trained workers and the United States cannot afford to fail to prepare its citizenry for that future. I maintain, however, absent an overhaul of the way we go about measuring achievement in science and engineering, making students of all races and both genders feel welcome in STEM training programs will not materially affect the country's ability to meet the technological and economic challenges of the future.

I urge two new approaches to fixing the evaluation system in science and engineering, one very detailed and suitable for personnel decisions and the other more appropriate for surveys of the gender gap in productivity. I have stated my views sharply with the hope my proposals unleash a torrent of fresh thinking on this issue.

New indicators of productivity: Ideas from the literature

A number of scholars have suggested alternative measures to the publications count: for example, Valian (1998) argued in behalf of using qualitative measures such as citations per paper. *Citations per paper* refers to the number of times the paper is cited in other scholars' work. Many studies have shown by this measure women do better, but fewer, studies than men.

Other scholars concerned with the deficiencies of the publications rate have suggested measures of scholarly performance such as academic leadership and quality of teaching as replacements. These scholars maintain holding office in professional societies, and editing professional journals also should be considered.

I have specific objections both to Valian's quality measure and to the suggestion of measuring other aspects of performance. As regards the latter approach, all of these measures are useful as tools for evaluation of academic staff. However, they are less relevant to the question of productivity and especially to the question I want to address: what measure(s) would be best overall to supplement the publications rate in studies of productivity such as those undertaken by Bozeman (2004), Xie and Shauman (2003) and other researchers interested in the issue of women's productivity relative to men's?

Valian has made a thoughtful attempt to supplement the publications count. However, as Valian herself conceded, measures of quality are controversial. People are much more likely to react negatively to invidious comparisons of quality than to a simple numerical count. Besides the drawback of being controversial, measures of quality such as citations per paper are difficult to compute.

I might nevertheless advocate for citations per paper if I were convinced it represents a valid quality measure. However, I believe "citations per paper" is an ambiguous measure of quality. It is possible to say something provocative but absurd and be quoted by any number of people irritated by your comments. A high number of citations per paper is not necessarily a measure of quality.

While I am unenthusiastic about "citations per paper", I applaud Valian's effort to find an alternative to the publications rate. Regrettably, too few such efforts have been made. We urgently need to develop productivity measures more revealing than the publications rate. We should concentrate on identifying additional measures of productivity rather than relying on a single replacement for publications. These supplements could throw additional light on the issue of scientific personnel performance without arousing the resistance trying to replace publications entirely as a metric might engender given how many people are committed to this familiar and appealing metric.

Della Porta, Triolo, and Fisher (2005), a revised version of whose work appears as an Appendix to this essay, have suggested assessing productivity in science much as it has been done in program evaluation and public administration research. A well-known model in those research traditions maintains productivity has two dimensions: effectiveness and efficiency. Effectiveness "equals results" defined as "throughputs plus outputs and outcomes compared to stakeholder expectations" while efficiency equals results divided by inputs. (*N.B.* In the revised version of the Della Porta, Triolo, and Fisher paper included as an Appendix, I have substituted goal attainment and externalities for "effectiveness" because that term has been used in so many different ways by various authors).

The approach advocated by Della Porta et al. is to assess the *research,* not the researcher. Their approach addresses some important defects of the quantity measure of productivity. First, it largely if not entirely eliminates the problem of rater bias in scoring men's work versus women's work (see Handelsman et al. 2005) because while the products to which the candidate contributed are assessed the candidate being assessed is not identified. Secondly, their technique eliminates another source of gender bias in the quantity measure of publications that stems from the custom of adding as an author the researcher whose grant funds financed the research even if that researcher has made no other contribution to the project. "Rainmakers" should be honored for their contributions—after all without the funds there would be no research to report—but the quantity measure greatly exaggerates their productivity. By the same token, the quantity measure approach shortchanges women scholars who do not have equal access to the funding sources yet manage to produce good work anyway.

Just as the new approach diminishes the productivity score of rainmakers, it increases the productivity score of the person who works on a project with little outside assistance and makes a useful contribution because it awards those persons points on effectiveness and efficiency.

Despite many attractive qualities to the approach Della Porta et al. propose including technical sophistication and its ability to address many defects of the simple quantity measure, its use will be primarily in personnel assessments because it is labor intensive. It may find use, for example, in situations where organizations are debating whether to award someone a major prize, or a big promotion such as an endowed chair. The major contribution of their proposal may well be to stimulate new work in this area.

A simple measure of productivity

A simple measure of productivity is still desirable for many applications such as evaluating how women are faring at the societal level compared to men. Although, the quantity measure has been used for many decades in this way, it has too many defects in its simplest form to merit continued use even for this purpose.

This is true even of the "short term" productivity measure advocated by Xie and Shauman (2003). They remark "the use of short-term measures is preferred for studies of sex differences in research productivity" and cite three reasons for this. First, they assert "women have only recently increased their participation in science and, therefore, have fewer years of experience than men on average." Second, they believe it is "highly plausible . . . women withdraw temporarily from active research" as a result of "spousal/child rearing constraints." The third reason they give is that

> it is difficult to incorporate explanatory variables measuring resource availability into multivariate models when the cumulative count of productivity is the outcome variable, because such explanatory variables are more likely to be endogenous rather than exogenous to one's cumulative productivity (2003:180).

The first reason, on closer examination, seems to be largely based on the influence of one source (Astin 1969). It really would not apply to more recent data. For example, the participation rate of women in physics has not measurably changed in several decades at the doctoral rate and the same applies to doctoral level engineers.

The second reason is also challenged by studies showing women scientists do not stop producing studies. Granted, anecdotal studies tell of women who balanced a phone in one hand and diapers to change the baby in the other. Somehow these women managed to continue grant seeking and research while also being attentive mothers. Of course, other women who were successful in science did not get married nor have children. However, they may not have wished for either even if they found eligible men willing to share the work of raising a family while pursuing busy professional careers.

The third reason seems to have the most substance to it. However, even here the real nub of the issue is the particular variables Xie and Shauman saw as pertinent explanatory variables (see Table 7, 2003:47). Indeed, with the explanatory variables they picked, the cumulative measure might seem inferior.

The short-term productivity measure, however, fails for the same basic reason as the cumulative measure more commonly used: it is, to an even greater degree, unable to handle the irrelevant response bias problem that throws its validity into question.

This, of course, creates a quandary of sorts. For some purposes, Xie and Shauman's measure is preferable to the cumulative approach to productivity. Some kind of measure combining relative simplicity of calculation with greater reliability needs to be developed.

Why do we need to supplement the publications rate?

Perhaps the best measure of the overall progress of women's productivity in the area of productivity is an index consisting of the simple publications count and one (or more) other simple outcome measures to supplement it.

A simple measure is needed for survey work

A simple measure of our nation's scientific productivity that, like the Dow Jones Stock Index, enjoys widespread public acceptance will always be needed even if it provides ambiguous data. Thus, while the publications rate is not a good metric (whether we are referring to a cumulative or short-term rate), for the foreseeable future, tossing out the publications metric is not an option. For most people a bad measure is preferable to none at all. Besides, tossing out the publications metric would not change the complexion of our nation's major research centers' senior faculty.

A second independent indicator can overcome the reliability problem

Unreliability is the principal objection to the quantity measure of productivity. Because it is such a crude indicator of resources, it is difficult to interpret. However, the short term quantity measure is superior in some ways (see Xie and Shauman) to the cumulative quantity measure.

The main hurdle to overcome is the irrelevant response bias possible in the short-term quantity measure. This threat to validity arises because systematic changes in ways women or men produce papers might irrelevantly influence the quantity count causing the relationship of the measured indicator to the real phenomenon of productivity to rise or fall unpredictably.

One of Ivey's suggestions to improve women's measured productivity illustrates the irrelevant response bias problem well. She calls for women to do more papers on the same project than they often do. For instance, instead of women scientists writing one report after the project is completed, she suggests perhaps the researchers do a paper on the methods of research being used in the study, another paper on the preliminary findings, and perhaps a final report with all the findings, including corrections of errors in the paper on preliminary findings. Thus, the writers would get credit for three publications instead of just one.

Although producing extra reports amounts to little more than resume padding, investigators do this all the time to squeeze as much as possible out of their projects to enhance their publications records

I oppose any practice as a bad idea that increases the unreliability of the publications count as would occur if Ivey's suggestion were widely adopted by women scientists. Rather we should seek better measures that more accurately portray women scientists' contributions.

One way to improve measures is to use multiple metrics of the phenomenon, that is, indices instead of a simple measure. Each component of the index may have advantages and disadvantages but the index measures the underlying phenomenon better than the single measure can.

Multiple measures are especially attractive when a single measure is vulnerable to irrelevant response bias. One frequently employed means to overcome irrelevant response bias might be to have an independent indicator, besides the publication count, that correlates highly with productivity, but is not highly intercorrelated with other predictors of productivity such as gender, a demanding but not impossible standard to meet.

A new perspective on productivity: Theoretical foundations

Any proposed measures of productivity must not be gender biased. Implementing biased measures favoring women leads to reverse discrimination. That is not only morally objectionable; it is likely to have only short-lived effects in redressing the lopsided ratio of men to women senior faculty. Worse, ultimately it will make entry to the top ranks of science and engineering even more difficult for capable women and minority scientists after the inevitable outcry.

If we were to design a productivity measure from scratch at a minimum it should have certain features:

(1) Simple to compute;
(2) Have "face" validity as a measure of productivity; and
(3) Be of policy significance.

Based on these criteria the following nine variables show promise as possible indicators:

1. Organization sponsor picked the problem;
2. Organization sponsor wants the political sensitivity of the issue considered in problem choice;
3. Researcher considers the social significance of the problem before selecting it for research;
4. Researcher considers whether the research can be completed within applicable deadlines before selecting the problem for research;
5. Researcher considers whether a renowned colleague suggested research on the problem before working on it;
6. Researcher prefers problems that others also are investigating*;
7. Researcher desires to publish in a new dynamic area*;
8. Researcher did not choose a problem whose attraction was that it would allow him/her to work with a technically superior colleague*; and
9. Researcher claimed to have a "fundamental value" research orientation. (See the definition for "fundamental value" orientation below, p. 70)

Of the nine variables in the above list, three (no.6-8 inclusive) are starred because they were predictors in the best (and all other) models considered in the multiple regression analysis presented in Chapter 6 of *The Research Productivity of Scientists* (hereafter, Fisher 2005). However, the first five variables seemed to be possible predictors in cross tabulations.

The starred variables may need a bit of explaining. Number six, for example, does not contradict number seven though it may look that way.

Number six suggests only that the researcher look for clues to what is an exciting area to work in by culling intelligence on what other prominent scholars and scientists are currently interested in doing. Understood this way, it is not really a very different variable from number seven in the list, perhaps only a somewhat differently worded variation on the same theme. Number eight is very different, however, and suggests smart scientists are wary of the "Mathew effect" whereby, as Merton (1973) showed, celebrated scientists get most of the credit from joint work with others, and avoid apprenticing themselves, except in special circumstances, once they have received their doctorates.

The paramount role of the organization in problem choice and productivity

Although the quantity measure of productivity is supposed to measure a scientist's output, Fisher (2005) in fact, showed that it measures the outcome of an organization process in which the scientist participates but does not have control over key aspects. For example, his study found that "the work organization loosens the researcher's commitment" to the paradigm(s) she came with while socializing her to embrace other paradigms the organization endorses. Typically this took the form of employing scientists whose degrees were all or mostly in one discipline on projects requiring a broader spectrum of technical knowledge (e.g. psychologists doing chemistry research).

Furthermore, the work organization socializes the scientist to accept other criteria of a problem's worth (e.g. contributes to "bottom line") beyond the mathematical and technical criteria that her discipline has taught her to apply (Fisher 2005:179). These conclusions are illustrated by the finding that 13 percent of respondents claiming the organization wanted them to do research of commercial value used theory from their own field to suggest a research problem suitable for investigation versus 46 percent who found their problem from other sources (Table 5.10, 2005:181).

> It might help the reader visualize the process of organization problem choice, in which the scientist is a participant but not the person in charge, to consider a hypothetical example. "Beltway Consulting" has a large contract to develop cameras for security around sensitive installations and it has hired a physicist to help develop these cameras. The company, in addition to wanting the physicist's expertise in optics, also wants the new hire to become familiar with social science theories of deterrence and social control. Furthermore, they want the physicist to embrace the company's interest in new cameras and other sensors because of the possibility of patents. Perhaps this physicist did her dissertation on new ways to measure the distance of extremely distant objects in space using satellite-based instrumentation. However, she could not find work in that

astrophysics specialty and turned to technology for criminal justice and military security as a field in which to earn a living.

In the above example, we see the company urging the physicist to work on applications of her optics training to criminal justice and security work. She has no academic background in such fields (her academic training and interests have been in optics and astrophysics) and the company needs to orient her to relevant literature so she works on equipment of value to them as a commercial enterprise. For purposes of this example, imagine the people who hired her are primarily men and their new hire is a woman. The hypothesis advanced here is that, consciously or not, her superiors will scrutinize her performance more closely than they would a male physicist to see that she really "gets it" and performs to the level they require. The possibility they will meddle in unconstructive ways, based on the ideas and findings discussed above, is also heightened because the scientist is a woman and the line supervisors are overwhelmingly male.

The control the organization exercises may vary depending on characteristics such as the gender of the scientist. This is dramatically illustrated by Fisher's finding that 39 percent of the women respondents compared to only 16 percent of the men ever had to significantly alter or abandon a study because of institutional controls (2005:205).

The close control of women in science and engineering suggested by the finding of differences between the genders in experiences with institutional controls suggests men supervisors have treated women in science as less trustworthy, less able to make decisions without supervisory input, and thereby have communicated women are really not welcome in this "man's world" of science and engineering. If this hypothesis is correct, then perhaps there might be differences among disciplines in the degree to which women reported needing to change or abandon a study because of institutional controls.

To assess whether differences in disciplines could be found, the data set used in my previous book (Fisher 2005) was reanalyzed. Table 8 shows the results of an analysis comparing men and women in STEM disciplines *other than* biology/medical research; fields where women have made strides in penetrating the top ranks of the discipline compared to all other STEM disciplines.

Forty-two percent of the women compared to 8 percent of the men in these non-biology disciplines (e.g. physics, engineering, behavioral sciences) reported ever having to significantly alter or abandon a project because of institutional controls. These results were highly significant statistically in spite of the small numbers of cases available.

Table 8: Ever had to Alter or Drop Study by Gender: All STEM Disciplines, Except Biologists and Physicians

Ever Had to Alter or Drop Study		
	Women	**Men**
Yes	42% (10)	8% (3)
No	58% (14)	92% (34)
X2 (with continuity correction) = 7.877, p=.005 Fisher's Exact Test (2 sided), p=.003		

Engineering, chemistry, and physics are of particular interest because they are disciplines where the number of senior women faculty is especially low. Did higher proportions of women in these disciplines report having to alter or abandon studies than in other non-biology fields?

In chemistry and physics, fields with relatively few senior women faculty, 60 percent of the women versus just 8 percent of the men respondents reported having to forego or alter a study because of institutional controls. Despite the small number of respondents the relationship was statistically significant. (X^2=5.716, P=.017, P=.044, Fisher's one-sided Exact Test). (Although the results with engineering included with chemistry and physics were quite similar, they were not significant at the .05 level in the analysis).

Assuming the difference found between men and women researchers is real, how would we interpret the finding? The problems with institutional controls that women encounter especially in physics and chemistry (and perhaps other disciplines) could be *both* cause and effect of their limited representation in the senior echelons of the university faculty. The effect of few senior women faculty in these fields, for example, is university administrators and male professors hostile to women's entry into these fields and other "dispassionate" sciences could introduce institutional controls or differentially enforce controls that adversely affect the productivity of women scientists. This would in turn be a cause of women's slow progress in gaining entry into the senior ranks in these disciplines.

Having emphasized that a problem choice in science is an organizational decision and that scientists do not control key aspects of the decision, it is now time to provide a more analytical view of the problem choice process in sociological terms.

The nature of the organization decision to study a problem

The choice of a research problem for study is an organization decision, not a scientist decision—though the scientist is (usually) a participant. However, the research problem choice is not a decision in quite the same way as a vote for a candidate in an election is a decision. A research problem choice emerges over time under conditions of ambiguity about essential matters. More accurately, the problem choice (1) is an organizational decision, (2) occurs under conditions of ambiguity, and (3) is the result of processes that have underlying regularities even if the choice itself is not predictable.

This view first outlined in 1990 by a Dutch scholar, Sjerp Zeldenrust, in his doctoral dissertation represents a sharp departure from the mainstream view that problem choices are individual scientist choices as exemplified in the work of rationalist scholars such as J. Cole and Zuckerman. However, some of the main features of the new conception were visible in early work in the sociology of professions and organization studies.

Other social scientists had also noted that scientists were not autonomous (in the sense of being only accountable to others in their discipline) in their problem choices. Shenhav (1985) had emphasized because scientists had a resource dependency on nonscientists who controlled resources needed for research the nonscientists could influence the problem choices. As far back as 1951 Talcott Parsons pointed out scientists were not truly autonomous but they needed some degree of autonomy in order to function in their professional roles (1951:335).

The "problem choice process" is the set of organizational processes that result in a decision about what problem will be investigated by scientists employed by or contracting with a sponsoring organization. The problem choice that results from this process was a linkage among three independent or loosely coupled flows: a demand (from a research sponsor) a problem (proposed by the scientist) and constraints (a separate flow that sometimes affects the linkage independently).

The choice occurs in ambiguous circumstances since ambiguity can be present in each of the flows. The ambiguity can take many forms: (1) potential sponsors may express an interest in the outcome of a particular piece of research, but not be willing to provide resources to do the study (ambiguity in the demand flow); (2) the sponsor may be willing to fund the study, but the technology to do the research validly does not yet exist (ambiguity in the problem flow); or (3) the technology that does exist is too costly to be commercially useful (constraint flow). Furthermore, developing technology to overcome the cost constraint may take more time than the sponsor wants to wait (and perhaps cost more in resources than the sponsor is ready to spend).

Whenever there is a problem in one of the flows resulting in a failure to have a linkage, it may unleash a process called "unblackboxing." "Unblackboxing"—Zeldenrust's coinage—is an unlovely but necessary term for a search process undertaken to permit the linkage to occur. Perhaps an illustrative example will help clarify this term's meaning. Suppose there is demand for cheaper petroleum. It is known oil sands in Alberta, Canada contain huge reserves of petroleum but using current technologies for extraction have made it too expensive until recently. However, with oil prices shooting higher those technologies are commercially attractive now.

Some firms in the oil business may still not be comfortable with the cost of the current technologies for extracting oil from the oil sands. A major investment up front is necessary and if the price of crude oil drops below a certain threshold the investment would be wasted. They are, however, willing to research new technologies that would be able to extract oil at a lower cost than the current methods. Scientists offering problems (the results of which may be new technologies for extracting the oil from the sand that are less expensive or that actually work in the real world rather than only in a laboratory setting) are "unblackboxing" the "problem flow" since the demand is already there for effective technologies and the constraint, an extraction cost below a particular threshold (e.g. $55.00 a barrel) is relatively constant (in real dollar terms).

Consider each of the flows. Demand is not constant. It can and does change over time. For example, a new cheap method for making clean diesel fuel from coal might cause companies to lose interest in extracting oil from the oil sands of Alberta and focus on acquiring coal deposits in Montana instead. The point is that changes in demand, after a linkage among demand, problem and constraint have formed, can cause the linkage to dissolve.

Just as demand flows change over time, problem flows and constraint flows also can change over time. Sometimes after a problem is offered, for example, research done elsewhere shows the problem to be absurd or obviates it. This, too, can disrupt a linkage. Sometimes a constraint no longer seems so Herculean. For instance, returning to our example of oil extraction technology, when oil was selling for 30 dollars a barrel, known methods to extract oil from oil-bearing sands might have seemed prohibitively expensive. When oil started selling at $66.00 a barrel, applied engineering work to perfect methods of extracting petroleum from oil bearing sands might have become quite attractive and work might commence. The same constraint no longer leads to either delaying research or hunting for very cheap methods of extracting oil from oil-bearing sands.

New linkages are forming all the time. In the example above, once the demand wilted and the linkage dissolved, a new demand fueled by an unexpected spurt in the price of gasoline could lead to a new linkage. In other words, after a linkage dissolves, new linkages can occur subsequently between a new demand, problem, and constraint. The new linkages, like the prior ones, occur in conditions of ambiguity and changes in the flows can likewise disrupt them. In abstract terms, Fisher (2005) argues changes could occur in any of these flows either singly or together (e.g. a demand and a constraint could change) with the consequence being the linkage could dissolve. A new linkage might then form quickly or slowly—resulting in a new problem choice—or not at all.

A problem choice is clearly a fragile thing since all kinds of events can disrupt it. For example, an organization culture that implicitly devalues work by women, or that sees women's work as more prone to error compared to men's can lead to closer scrutiny of women researcher's projects and more frequent alterations or cancellation of those projects as was shown above. When asked if they ever had to terminate or significantly alter a study because of "institutional controls", 39 percent of the women respondents stated that this had indeed occurred versus 16 percent of the men (Fisher 2005:205). Why women should have experienced these events more than men—that is, an apparent *gender patterning*—needs to be pursued in future research.

Not only do these findings indicate the problem finding and research process for women in the STEM disciplines is often more arduous, they also point to reasons why women produce fewer publications than men. And they underline the fact that problem choices resulted from organization decision making processes similar to the decision making processes in universities as described by James March and his collaborators (see e.g. Cohen, March, and Olson 1972). Far from being a rational choice of scientists fully aware of the intellectual significance of the problem they were interested in investigating, the choice of research problems results from the interaction of many persons, including nonscientists, and occurs under ambiguous conditions.

Only beginning with Fisher (2005) has sociological research begun to investigate the link between the problem-choice process and productivity (as measured by publications). This is partly because the main paradigms followed by sociologists in research on problem choice and productivity focused on the individual scientist, rather than the organizational level, and failed to note the link between the process and the resulting productivity. For example, rationalist scholars understood organizations could influence scientist behavior, but largely viewed this influence negatively, as interference in work scientists are best equipped by training and familiarity with the paradigms of their discipline to do. Rationalists presumed scientists knew ahead of time the intellectual importance of a problem they sought to study—a view challenged in Fisher (2005). The main critics of the rationalist view, loosely termed "social constructivist scholars", denied scientists ever derived problems from theories but rather from prior findings. While they were correct in criticizing the rationalists for exaggerating the importance of deductive research, the social constructivists erred in exaggerating the importance of inductive research. In sum, whereas the rationalists saw the role of the organization as essentially negative, the social constructivists saw it as inevitable. Neither realized sometimes problem choices could be inductive and sometimes deductive depending in part on whether the intellectual significance of the problem was known in advance and whether the organization's role in the problem choice process was meddlesome or constructive.

Given the pervasive influence of the organization sponsor, the difference in the problem choice processes of men and women is highly important since it suggests the women experience the influence of the organization differently. Perhaps, for example, the organization sponsor feels women cannot be trusted as much as men to supervise a staff of junior researchers and interfere more with the work of teams headed by women senior investigators, reducing their output. Perhaps, since women are more dependent on government grants, they experience interference more often, not because they are women, but because proportionally more of them than men experience the micro-managing proclivities of government grant monitors. Other reasons may account for women than men reporting that they had to alter or cancel a project because of institutional controls. Future research may point to these additional factors.

Perhaps because of their greater probability of having their research cancelled or significantly altered as a result of institutional controls women have adopted certain patterns of research problem choice and research style quite different from men. According to Fisher, women, far more than men, pay attention to applicable time deadlines in planning their studies; women are more likely to consider if a renowned researcher has recommended research on a subject than men are; and women are more likely to consider if the problem is related to public concerns or controversies (Table 5.19, 2005:202-203).

It is possible women's being especially meticulous in planning and carrying out their research is a consequence of exaggerated fears of harassment based either on rumors or on a previous experience. Whatever its causes, these gender differences in planning and executing studies are potentially significant variables in trying to understand the gender publications gap.

What makes a phenomenon a good indicator of research productivity?

Finding an indicator is a lot simpler than finding a "cause" of a phenomenon of interest. It is not necessary for an indicator of a phenomenon to be causally related to the phenomenon. It merely needs to be correlated reasonably highly with the phenomenon. (A spurious correlation will do, in fact, for an indicator though clearly not for a causal connection). For example, musical talent may be correlated with ability to do well in architecture. Why a fine musical talent is a predictor for success at completing an architecture degree program in college is not readily apparent. (Some educators speculate that the genes responsible for musical ability and architectural aptitude may often occur together) However, given intense competition for entry into an architecture program, the school could reasonably favor musically gifted students over others who are not musically gifted when deciding among candidates similar in grades and other clearly relevant factors.

Based on the criteria of correlation with the phenomenon of interest, I believe the indicators given above, and possibly others like them, i.e., problem choice process and research style variables, are good candidates as indicators. My argument rests on a hypothesis first advanced in *The Research Productivity of Scientists* (2005) that the differences in the problem choice process that men and women scientists follow accounts for their differences in productivity.

Rationale for variables

Analyses of the relationship of variable 9 to productivity were not presented in *The Research Productivity of Scientists*. However, this variable shows some promise as an indicator, at least for one important component population of STEM disciplinarians.

Research orientation refers to the style the researcher follows in finding problems for research. Three styles were identified: growth, fundamental value, and eclectic. Growth-oriented researchers are opportunistic, keeping an eye on and moving into dynamic areas, that is, those attracting numerous capable scientists because good problems are readily available and leaving areas where few good research problems remain.

Fundamental value researchers, on the other hand, are willing to look at virgin territory, problems neglected for some reason, but which these scientists see as potentially offering rich payoffs if someone would invest the necessary time and effort in researching them. Eclectic researchers would adopt either a growth or a fundamental value approach depending on circumstances.

In disciplines where significant numbers of women have penetrated the top tier (e.g. psychology) the proportions of men and women scientists with fundamental value orientations will be similar in contrast to fields where few women have succeeded in penetrating the top tier of the profession.

Overall, fewer women have a fundamental value orientation because (1) in many disciplines only limited numbers of women have penetrated the top tier of their profession. Furthermore, women face resource deficiencies in an environment that also demands publications as evidence of productivity. Therefore they may innovate by moving to growth orientations and/or eclectic orientations.

Findings

The first question is whether those who saw themselves as fundamental value researchers really were different in meaningful ways from the other types of scientists. The data to answer this and all other questions in this section came from the data set used in *The Research Productivity of Scientists*, a non-probability sample survey of 107 scientists from several scientific disciplines.

Generally, there was no relationship between type of research orientation and research planning styles that might influence productivity. However, among biologists/medical researchers the data indicated fundamental value researchers go about the process of finding worthwhile problems differently than do growth or eclectic scientists. For example, Table 9 shows before choosing a problem to investigate, the fundamental value scientists in biology/medical science are far more likely (80%) to consider whether they needed to borrow analytical techniques from other fields to solve problems in their own field than are scientists who were "growth oriented" or "eclectic" in style (42%). The fundamental value scientists were also more likely (86%) to read widely outside their own specialty before choosing their research problem than were the researchers whose orientations were growth or eclectic (42%). Furthermore, before choosing their research problem they were also more likely (80%) to talk to colleagues in other fields than were scientists who did not have a fundamental value orientation (37%). These differences between fundamental value scientists and the others suggest that fundamental value scientists spend more time in the problem finding process than do scientists with other orientations.

Table 9: Research Planning Steps by Type of Research Orientation (Biologists/Physicians Only)

Research Planning Steps by Research Orientation (Biologists/Physicians Only)		
(a) Borrow Research Technique?	**Fundamental Value Orientation**	**All Other Orientations**
Yes	80% (12)	42% (8)
No	20% (3)	58% (11)
Fisher's Exact Test (2 sided) p=.038		
(b) Talk with Colleagues Outside of Field?	**Fundamental Value Orientation**	**All Other Orientations**
Yes	80% (12)	37% (7)
No	20% (3)	63% (12)
Fisher's Exact Test (2 sided) p=.017		

The next question was whether *gender differences* existed in the proportions of scientists with the respective orientations. No such differences might indicate men and women scholars were equally able to count on long-term support. Is the proportional distribution of women and men across research orientations essentially similar, suggesting resource differentials were not constraining women's research problem choices more than men's?

The results of the analysis were suggestive but not conclusive. Overall, men and women scientists described themselves as fundamental value oriented in roughly similar proportions. However, in biology/medical research where major strides are being made in the fight against serious diseases, the results were rather different. Fifty-seven percent (57.1%) of the men biologist/medical researchers versus only 29 percent of the women claimed to have a fundamental value orientation. The result was only marginally statistically significant (p= .094), possibly because of the small numbers of cases available for the analysis.

Denied some of the resources they seek, women scientists may find the level of support they receive forecloses investigating some kinds of promising studies. That forces them to do, perhaps, less interesting research. Lack of long term support means that more time is spent in the hunt for funds and less is spent on research itself, perhaps elongating their problem finding. Greater interference in their research from officious sponsors may result in more abandoned or altered projects for them than for men. All of these could slow women scientists' progress in reaching the senior ranks in the professions and especially in the academic world. These kinds of problems may be behind the difficulty women are having in engineering where women's progress has been slower than in all the other academic professions.

Although the possibility of gender based resource differentials in biological/medical research is troubling in itself, it is also important to assess what its effect has been on productivity. Did researchers with one type of research orientation have significantly higher cumulative productivity than those with the other types, regardless of gender?

Do those who followed a fundamental value approach have a higher level of cumulative productivity than those with other orientations? The data did not show this to be true. However, about twice as many scientists of all respondents in the study population with a value orientation (37.5%) were in the highest quartile of productivity compared to 19% of those with an eclectic or growth orientation. While it is not necessary to have a value orientation to be a very productive scientist in purely quantitative terms it probably was beneficial in some significant fashion. Otherwise, why would far more of the fundamental value oriented scientists than of the others be in the top productivity group?

A value orientation is primarily relevant to the kinds of problems on which the scientist would consider. That fewer women than men in biology/medical science have a value orientation is an indication, not that women care less about doing fundamental research than men, but that given the relatively scant resources for these women researchers they must be opportunistic, willing to do papers feasible within the resource constraints under which operate. Quite simply, for many women in biology/medical research, a fundamental value orientation is not an available option. They either must be growth oriented or eclectic in their approach to research.

At least for now the mere fact that women biologists cannot easily adopt a fundamental value orientation does not prevent them from doing a large number of research papers. Eclectic as well as fundamental value scientists are among the top scientists as measured by the number of publications. However, if the research requires significant long-term financial backing, the other research orientations are not as promising if the aim is to do important work leading to breakthroughs, and not simply a high volume of papers. This means gender differences in percentages of scientists exhibiting a value orientation may point to a serious problem for American science: Too few women can count on the kind of financial support needed to make breakthroughs. Simple publication counts do not reveal this problem. The best these measures can do is to show an adverse effect on *publication* rates from resource deficiencies as indicated by Xie and Shauman (2003) and Fisher (2005).

What negative consequences do women suffer because of these kinds of assaults on their dignity and competence? It may discourage talented women from staying in the field. Among the women who choose to stay, however, what are the consequences of the harassment? Does it reduce their productivity? Certainly it does not reduce their *publications* rate.

However, given the pressure to publish quickly exerted by both their academic departments and government sponsors (on whom women are more dependent than men for their funding) and whose research agenda shifts from year to year, women scientists may find it more difficult to plan and execute intellectually challenging work than their male counterparts. Supporting this are (1) Stephan's observation about grant supported research; and (2) data adduced about the characteristics of women scientists that reduce their publications e.g. their paying close attention to deadlines (see Fisher 2005:203) that probably reflects their greater exposure to government funding.

Elizabeth Ivey, the president of the Association of Women in Science has pushed the idea of "publish earlier and more often" as a trick of the trade that women need to learn. "Women researchers don't tend to publish their results until they're very near the end of their project, whereas male researchers will publish intermediate results all along the way, so they build maybe three articles on a project where women tend to have only one" according to Ivey. Ivey's observations are consistent with Fisher's findings women are more meticulous in planning and executing studies, perhaps because they are concerned they will be scrutinized more carefully than men.

Ivey hopes to level the playing field for women with a "quick fix" that, unfortunately, may backfire. Suppose Ivey is successful in her plans to teach women how to generate more publications and the peer review system responds by counting each woman's publication as less than a man's. This would negate whatever benefit Ivey hopes to accomplish with her proposal. Remember Wenneras and Wold showed women needed to do 2.6 times more publications than men to receive a grant from the Swedish Medical Research Council. Furthermore, will embracing Ivey's idea not undercut the argument we need to develop a valid new conception of productivity reflecting the real scientific contributions of all scientists? Ivey's suggestion lends legitimacy to an unreliable measure of productivity.

Summary and conclusions

The need is urgent to develop valid new productivity measures. Women's progress in penetrating the upper ranks of the STEM disciplines, as well as the progress of under represented minorities, depends on having adequate tools for appraisal of their scientific contributions. Two promising approaches were discussed in this chapter. One is a model of productivity based on work in the program evaluation and public administration research traditions that might find use in appraisals of particular scientists' contributions. The second is a tool better suited to survey studies of comparative productivity of different component populations such as men and women scientists.

Developing new metrics of scientific productivity is a major challenge. This chapter only hinted at the difficulties that must be addressed. For example, researchers developing tools for scientist appraisal will need to be cognizant of the important consequences for their work of disciplinary differences. The findings presented in Table 8, in respect to the gender differences with institutional controls forcing studies to be altered or terminated, highlight the difficulties women in physics and chemistry, and probably engineering as well, have in building a track record of completed projects on which their productivity can be fairly assessed.

On the other hand, in biology and medicine, where the percentage of women scientists is greater than in physics, women scientists may be able to produce similar quantities of papers to men colleagues. However, with more women dependent on federal grants with annual funding cycles, fewer women biologists/medical researchers can afford to have fundamental value orientation that offers the best chance to do really important groundbreaking research. Women in biology/medical research are more likely than men to follow a growth or eclectic orientation migrating into fields that are becoming "hot" and getting out of specialties that seem "played out." Therefore, in biology/medicine, assessments of women's productivity may need to look at changes over time in types of funding, from annual funding cycles to longer term cycles, and its impact on the intellectual significance of women scientists' research.

Chapter Four

Conclusions and policy recommendations

As a society that believes in the right of every U.S. citizen to liberty and the pursuit of happiness, ideas enshrined in the documents on which our nation was founded, we Americans are concerned about fairness in every sector of the nation's economy. Science and engineering, however, are of special concern. Our nation became great because of the numerous inventions and discoveries of our scientists and engineers, e.g. the electric light bulb, the telephone, the airplane, television, computers and vaccines and medicines for the most dreaded diseases of mankind such as polio, heart disease, cancer, and other crippling or lethal illnesses to name just a few significant contributions.

The importance of science and the fruits of scientific research to the national health, safety and prosperity were recognized officially by the United States Government in 1945 when Dr. Vannevar Bush issued his famous report, *Science: the Endless Frontier*. That report underpins United States Government policy to this day.

Bush famously and farsightedly called for recruiting for scientific careers people of great promise and interest in science from every ethnic group, both genders, and without regard to income level, national origin or religion. The only test was to be ability, demonstrated through performance.

Our educational system has not lived up to this promise. African American, Latino, and Native American peoples are grossly under represented in the ranks of American science and engineering, especially at the most senior levels. Women of all ethnic backgrounds, although present in significant numbers at the junior faculty levels and in the ranks of students at the undergraduate and graduate level, are also under represented in every rank among faculty, most especially at the most senior levels of science and engineering.

Our appalling failure to harness the talents of all our citizens in science and engineering is a national disgrace, at once a massive policy failure and a social science puzzle. We simply do not know why African American, Hispanic Americans, and Native Americans have not sought careers in science and engineering in significant numbers in the past nor do we know what to do to recruit more of them. Our lack of a national commitment to solve this problem is also a costly error we must rectify soon if we are to remain preeminent in science and engineering, keys to our economic prosperity and military security.

Recommendation: The country must commit to a major increase in the numbers of women and minorities in science and engineering and should create far more targeted scholarships for women and minorities to study these subjects at both the undergraduate and graduate levels.

Simply because we recognize that overcoming the problem of hostility to women in certain disciplines will not occur quickly no matter what we do, we cannot afford to be complacent or indifferent to the harm to our nation's economy and the injustice of the present situation. We can and must commit to making the climate in colleges and universities more welcoming to women and minorities. To this end we must implement a variety of new strategies for accommodating these under represented groups.

Recommendation: Special grants from government agencies and non-government foundations to colleges and universities to help women and minorities be successful in science and engineering studies should be widely available. Such grants would be made upon application by the schools for innovative strategies of mentoring, intensive pre-college summer programs, counseling programs and the like. Sensitivity training for professors, and firm written policies against sexual harassment or racial hostility must be promulgated where appropriate in engineering and science departments. High schools should be eligible for special grants programs to encourage them (a) to identify promising women and minority science and engineering prospects and (b) to counsel students with aptitude in mathematics and science aptitude to pursue science and engineering careers.

While the academy, government and industry must do more recruitment of under represented minorities and women at the entry level in science and engineering disciplines, much more progress in this regard has occurred than in the promotion of these groups into the higher ranks of the professions. The lopsided ratio of men to women and the near absence of under represented minorities in the top echelons of science and engineering will continue to prove a challenge to overcome and, unfortunately, it will be decades before the current programs to recruit more women and under represented minorities show up as significantly higher proportions of women and under represented minorities in the top ranks of scientists and engineers in the academic sphere.

Many reasons account for this. For instance, in some STEM fields, simply bringing in minorities and more women at the entry level will be a slow process even if our society makes the kind of commitment it needs to redress these problems in part because our understanding of the processes by which students choose their major fields of concentration in college is limited.

A second reason why the lopsided ratio of men to women and the near absence of under represented minorities in the top echelons of science and engineering will continue to prove a challenge to overcome is a relationship between promotions and other rewards on one hand, and productivity on the other, though it is not a perfect correlation by any means.

Americans have not yet accepted that women have the same potential to be good scientists and engineers as men do. In various ways, we shortchange women as well as minority scientists and engineers in terms of research resources, salary, promotions, and tenure resulting in such phenomena as the gender gap in publications. Redressing this imbalance in resources is necessary before the gender gap, much of which is a resource gap, will disappear and more women and minorities will be promoted into higher positions.

Third, any improvement in productivity of women and minority scientists and engineers as a consequence of more equitable distribution of resources will not translate into instant improvements in the ratio of women to men at the senior faculty level. The tenure system slows down any movement of women and under represented minorities into the senior ranks of scientists and engineers at our colleges and universities.

Fourth, Chapter 2 presented data indicating different disciplines have different male-to-female ratios of full professors, suggesting differing attitudes toward women's abilities as scientists and engineers across disciplines. For example, among the social forces the effects of which must be blunted are traditional male engineering culture as dissected by Cockburn and other feminist scholars. The male engineering culture has been impervious to evidence of women's capacity to do quality work in engineering. Until women move into the dispassionate sciences and engineering in large numbers, no change will occur in the attitude of many men professionals that women are outsiders who do not belong.

New conceptions of productivity are necessary as part of a comprehensive plan to restore American scientific prowess in science and engineering; without this change in our assessment process the other steps—recruiting more women and under represented minorities into science, and making our educational institutions more welcoming of them—are unlikely to succeed in bringing more women and under represented minorities into the top ranks of American science.

The alternative to developing new conceptions of productivity is continued reliance on recruiting foreign talent to staff positions in academy, government and industry. The drawbacks of this dependence on immigrants, even when we are successful in luring them here, make this an alarming prospect.

Chapter 2 discussed the need for new ways to measure productivity in science and engineering. The current measure quantity measure of productivity is unreliable and biased. Far from being a measure of productivity, it is part of the failed system of science education and selection of our nation's senior engineering and scientific leaders. Continued use of the simple quantity measure of publications perpetuates a white male dominated world in the STEM fields that will not produce sufficient scientists and engineers to meet this nation's needs. New ways must be found to measure productivity more reflective of women's contributions as scientists and engineers.

Recommendation: College and University faculty must evaluate their current appraisal systems for promotion, tenure and salary by funding outside consultants to perform assessments based on study of the current procedures, comparison with programs at other schools with relatively greater success in attracting and promoting women and minorities and making recommendations for modifications to make the evaluation procedures more gender neutral. All colleges and universities should grant maternity leave to men and women faculty automatically and provide day care for children of faculty and students at an affordable cost as part of their compensation programs and student services.

Chapter 3 presented two suggestions for new ways to measure productivity in science. In my opinion, the paramount and immediate task of the social science of science is to devise new ways to measure productivity in science. Without new conceptions of productivity to replace the simple measure of quantity the other steps—recruiting more women and under represented minorities into science, making our educational institutions more welcoming of women and under represented minorities—are likely to be unsuccessful in bringing more women and under represented minorities into the top ranks of American science.

The two suggestions that third chapter presented for new ways to measure productivity in science cannot be presumed inherently better than the current measures. They are only better if they address the defects of the present measures without substituting equally or more objectionable defects of their own. The two proposals for measuring productivity made in this essay are improvements over the simple quantity measure of publications in some respects but at the least they require assessment and perhaps further refinement.

Recommendation: The United States Government should fund the National Academy of Sciences or other appropriate group to do development work on and a detailed assessment of methods of appraising scientific and engineering productivity with the purpose of having such improved methods adopted by institutions of higher learning as part of their assessment procedures for tenure and promotion.

Appendix One

Measuring research productivity in a gender neutral way

Introduction[1]

Are women as suited as men to do scientific research? The answer to this question is an urgent policy issue. Demographic changes—fewer scientists coming to the United States, fewer men majoring in science and engineering than in the past, growing numbers of minority people in the American population—threaten to reduce the supply of scientific, technical, engineering, and mathematical (STEM) disciplinarians needed to maintain scientific and technical supremacy on which American economic prosperity and military prowess depends. These demographic pattern changes make it imperative we identify and harness the talents of people, e.g. women and minorities, from whose ranks few STEM professionals have traditionally come.

Recruiting more women and minorities into science will take major steps. Etzkowitz, Kemelgor, and Uzzi (2000), for example, shows the need to change graduate and undergraduate education; Rosser (2004), Xie and Shauman (2003), and Fisher (2005) among others show the chilly environments in which women work as scientists and engineers must be made more accommodating.

Part of that accommodation must be a rethinking of the way in which the academic world conceptualizes and measures scientific productivity. A number of authors (Valian 1998; Bozeman 2004; Della Porta et al. 2005 have identified problems in the traditional criteria for measuring academic performance adversely affecting women's measured productivity more than men's. Consequently, this paper maintains the academic evaluation system is biased against women though the bias may not be deliberate.

In the academic sphere the two primary measures of research performance are quantity and quality, according to Virginia Valian (1998). Quantity is typically measured as a simple ratio of "numbers of publications" divided by "a unit of time." However, as Valian notes, "the count is not completely straightforward" (1998:261). Valian observes a short article reviewing a book is a publication and so is a lengthy article reporting the results of a series of important experiments. Book reviews are publications just as are books. Valian reasonably asks "Are all articles equal? How many articles equal one book?"

Despite the defects of the publication count, it has been employed as the chief measure of scientific productivity in numerous studies for over forty years. Virtually all of those studies have found that men publish more research studies than women.

Some scholars (Mary Frank Fox 1991, 1995 and Xie and Schauman 1998, 2003) point to resource differentials between the genders to explain the gender publication count differential. Other scholars, notably J. Cole and Singer (1992), emphasize socialization differences and especially the different ways women react to disappointment and setbacks in their lives compared to men.

Until now, however, no one has challenged the conception of productivity as somehow equivalent to the publication count despite its well-documented weaknesses as an indicator. This paper's position is women STEM professionals will not make significant progress in achieving equality of opportunity with men until the evaluation criteria of professional accomplishment in science are reliable and gender neutral. Therefore, developing valid indicators of performance is an essential task if we are to recruit more women into careers in science and engineering. Work on these evaluative criteria must proceed alongside other policy initiatives to make science and engineering more attractive to women and minorities. The United States can no longer afford the morally indefensible posture women Hispanics and African Americans are not able to do first-rate science.

The concept of scientific productivity

Before turning to the details of how to compute the new measure of scientific productivity, the concept of productivity as it is used in various disciplines needs to be discussed.

Scientific productivity has come to mean publications, and specifically publications in a standard unit of time. This conception of productivity is unique to the sociology of science; in other fields the concept of "productivity" generally means something quite different. Economists traditionally have equated productivity with efficiency as in the amount or value of output produced per unit of input. Increasing economic productivity meant increasing the ratio of outputs to inputs.

Students of the service economy (see e.g. Hage 1984) shifted the focus from efficiency to goal attainment, that is, the extent to which products and/or services satisfied consumers. Hage coined the term "attribute productivity" to emphasize the subjective nature of goal attainment and also that it is measured against a shifting array of attributes that change over time.

The economic concept of "externalities" is also pertinent here to emphasize scientific research can have various measurable consequences such as new products, new technologies and other consequences.

A new measure of scientific productivity

This paper proposes a new measure of scientific productivity, assessing research goal attainment and externalities on one hand and the efficiency of the research against expectations of stakeholders on the other. The new measure primarily builds on work in the organization studies literature (e.g. "open systems design theory" of James Thompson 1967) and the program evaluation literature, especially non-experimental approaches such as multi-attribute utility analysis, goal attainment assessment and cost benefit analysis.

Writers in the open systems design tradition have advocated (1) examining the relationship of inputs and throughputs on research outputs and outcomes (see Balk 1990); (2) direct measures of research effectiveness (see Pressman and Wildavsky 1973); and have emphasized (3) there is no single overarching concept of research effectiveness (Van de Ven 1986). Measures of effectiveness are multi-dimensional and include consideration of short and long-term implications (see R. H. Red Owl 1995), competing values (see Quinn and Rohrbaugh 1983), and the existence of multiple stakeholders and disciplines (Senge 1990).

Economists studying externalities of project investments have also enriched the understanding of research effects. From an economics standpoint, new knowledge is a good with the unique characteristic of being "consumed" without the stock being reduced.

Of particular interest to economists interested in externalities are the effects of the research process and of the new knowledge in the economy. Economists have defined different kinds of externalities including "real" externalities and "pecuniary" externalities among others. Below, I shall have more to say about this issue as regards the effects of research and the research process from an economics standpoint.

The organization studies perspective of Thompson and others referred to above is consistent with new developments in evaluation research emphasizing project effectiveness is best considered as a revision of the prior probability of a result; effectiveness also must take into account projects have outcomes of interest to different stakeholders whose priorities differ, may conflict and compete. Techniques of evaluation research suitable for this approach to project effectiveness include Bayesian approaches (see Pollard 1986), multi-attribute utility analysis (see, for example, Edwards, Guttentag, and Snapper 1975) and cost-benefit analysis traditions (see e.g. Mishan 1976).

Conceptualizing productivity as a revised probability of goal attainment and also from the standpoint of externalities is a sharp departure from the conventional treatment of scientific productivity by writers in the sociology of science literature. Generally, by scientific productivity, sociological researchers have meant the number of publications (see Cole and Singer 1992; Valian 1998; Xie and Shauman 2003; Fisher 2005). Total productivity or "cumulative" productivity has sometimes been distinguished from "short term" productivity as in Xie and Shauman. However, what is common to all these writers is a simple identity: number of publications **(N)** divided by a standard unit of time **(T).**

Eq. 1 : $P = N/T$.

Following Fisher (2005), this paper maintains science is about the production of novelty of a particular type. It is novelty that has some intellectual value and perhaps some practical value as well (though this latter is not necessary for the result to be good science). The value of this novelty, in principle, can be assessed by scientists and by laymen, if the novelty also has consequences of a practical nature. The assessment of novelty by scientists is straightforward—it is a comparison of the results to the expectations. Mathematically this is equivalent to a revision in a prior probability of the result. The greater the revision in the prior probability the greater is the value of the novelty.

Eq. 2: $V = P_1 / P_0$

The questions answered by the new measure

The new measure represents a paradigm shift in evaluation of scientific productivity because not only is it a different concept, it is a different way of looking at productivity (see T. Kuhn 1962; 1970 on paradigm shifts). Paradigm shifts change the research questions and the new measure satisfies this criterion. For example, the new approach asks:

(1) Are any differences in research productivity by gender and/or ethnicity unrelated to numbers of publications, where productivity equals research goal attainment, externalities, and efficiency?
(2) If differences are found, are they in inputs (resources), throughputs (methods/processes), or outputs/outcomes?
(3) If differences are found, how are specific types of differences distributed across research areas, disciplines, or organizations?
(4) Are significant differences in research productivity found among women or ethnic groups within the same fields and types of organizations?
(5) If intra-gender differences are found for women, are they the same for men? Similarly if intra-group differences are found for specific ethnic groups, are they the same for the members of other groups?

Why a new measure of scientific productivity is needed now

Despite being an unreliable tool for assessing individual performance the quantity measure of productivity (Eq. 1) remains a major tool for this purpose. The simple quantity model of productivity achieved its popularity because it "appears" on the surface to be unbiased; furthermore, it is easy to compute and using it avoids the kind of disputes arising when attempts to measure "quality" are made.

However, the simple quantity measure victimizes women scientists because they have fewer collaborators than men (Bozeman 2004). This is particularly a problem in STEM disciplines since most research there is done in teams, enabling men to accumulate more publications than women scientists. Xie and Shauman also showed women have other resource deficits: proportionally more women scientists are assigned high teaching loads; fewer women scientists than men have private funding; and more women than men are dependent on government funding with its complex application procedures.

Fisher (2005) found women have experienced more interference and more constraints on their work than men. Valian showed women are promoted and achieve tenure more slowly than men.

Together with a propensity to credit celebrated scientists—most of whom are men—more than less famous scientists ("Mathew effect") for their scientific achievement, the quantity measure of productivity is a prominent feature of the *de facto* system of allocating rewards in science that has resulted in a lopsided ratio of men to women senior faculty members in the STEM disciplines. Continuing to legitimize the simple quantity measure of scientific performance rather than seeking other measures more reflective of the value contributed by researchers perpetuates this unjust system of reward allocation. It also supports the myth that when women and minority scientists do achieve high position it is merely because of their protected category status rather than because of their accomplishments. Assessing the novelty value of the research compared to expectations would be a way to overcome some of the problems of the simple quantity measure of productivity commonly used today.

The injustice of the current system of faculty recruitment and selection is not the only reason to drop the simple quantity measure of productivity in favor of alternative measures. New measures, such as proposed here, are needed urgently because the face of the workforce is changing. As noted elsewhere in this monograph, the current system of scientific education is not producing enough white male scientists to meet the needs of academic, industry, and government employers for doctoral level scientists, necessitating recruitment abroad. Importing talent from abroad is no longer an easy option because fewer S & E personnel with advanced degrees are immigrating to the United States to take jobs. Foreign talent can now find good opportunities in their own countries such as Taiwan, Singapore, India, and Brazil. In an increasing number of high technology spheres these countries are now competitors with the United States and our scientific and technical superiority; a key to our continued prosperity and military security is being eroded. We now have to make effective use of talent drawn from sectors of the native born population from whom we have only selected token numbers of scientific and engineering personnel in the past. That talent is available is sufficient numbers if the educational system and the academic work environments can adapt to accommodate them.

Computing the new measure

The new Measure relates inputs (I) and throughputs (T) compared (II) to stakeholder expectations (E_x):

Productivity (P) = Goal Attainment (G) and Externalities (E_t) and Efficiency (E) or ([G + E_t] + E)

Results (R) = Innovative Throughputs (T) plus Outputs or Outcomes (O) or (T + O)

Goal Attainment (G) = Results II Expectations (G= R II E_x)

Efficiency (E) = Results/Inputs (R/I) II Expectations or (E = [R/I] II E_x)

Components of the new measure are as follows:

Inputs: These are resources and environmental/organizational characteristics, drivers and constraints that may either propel or restrain investigation of a specific topic/question.

Throughputs: These are conceptual and methodological approaches that utilize resources in ways that add value/contribute to results. Throughputs can become research by-products since once documented they permit researchers to repeat experiments, confirm reliability, and validate findings. Innovative throughputs are counted as results because they contribute to the search for knowledge.

Outputs: These are the products and/ or research findings.

Outcomes: Influences on or implications for the future.

Results: These are intended *and* unintended outputs and outcomes that add value by furthering knowledge. Innovative throughputs that further future research may also be included. Results can also include negative externalities of the research such as environmental pollution, illness, etc.

Expectations: The standards and values that stakeholders within a specific subject area attribute to inputs, throughputs, outputs, and outcomes associated with a particular research project or area of study. Expectations characterize the constructs that are used to identify and assign weights to variables representing various components of the new model.

Stakeholders: Persons with significant knowledge, experience, and/or a interest in the research subject area. They may include scholars, students, practitioners, and consumers. The same or separate sets of stakeholders may choose, assign weighs, and/or score variables of the new measure in order to assess specific research projects/initiatives.

Goal Attainment: Results compared to stakeholder expectations (R II Ex). This can be assessed as a change in the prior probability of a particular outcome as a consequence of the project findings and outputs using well known techniques of multi-attribute utility analysis (On MAUA, see for example Edwards and Newman (1976). Externalities are relevant here as well and, as stated earlier, these can be both negative and positive. An infamous example of a study with negative externalities was the Tuskeegee study that looked at syphilis in African American men. Many people died for the sake of some information that could have been obtained without such an appalling disregard for human life.

Efficiency: The ratio of results to inputs or resources expended (R/I). Efficiency focuses on the amount/value of resources that were needed to produce desired outputs/outcomes. Efficiency helps determine whether the results of a particular research project were worth the investment when compared to alternative investments and deferred opportunities.

Efficiency as a criterion in evaluating a specific research project can be of virtually no importance to singularly important. Research projects have cost constraints and goal attainment is not only the accomplishing of the main objectives of the study but also its accomplishment within the cost constraints. Sometimes a study can only be accomplished when researchers figure out how to piggyback it onto another. For example, much social survey research on discrimination in the 1940s and perhaps afterward occurred as a result of such piggybacking of the survey questions about discrimination onto surveys undertaken for market research purposes of commercial clients. Instances where researchers ingeniously "bootleg" one study onto another (on "bootlegging" see LaPorte and Wood 1970) certainly exemplify the case when efficiency of the research should be noted in appraising scientific work.

Operationalizing the new measure

The new measure may be used to assess the productivity of an individual research project or multiple research projects or initiatives in the same area of study against a conceptual construct that uses a common set of weighted variables.

Assessing the productivity of a single project/initiative

The first step in the process is when experts and other knowledgeable stakeholders in the subject area identify significant input, throughput, output, and outcome variables which are used to assess the productivity of research as reported in publicly available publications as follows. Then, the experts assign weights (absolute or relative) to variables consistent with stakeholder values and expectations for the area of study/topic.

Third the experts review publicly available reports/publications, within a specific timeframe, in date order against the selected variables and score each variable once, based on the total body of work reviewed.

The next step largely is determined by the conventions of the discipline. The simplest procedure would be to add the scores within a category and arrive at a cumulative score for inputs, throughputs, and outputs/outcomes. However, the scores are values on ordinal scales, not pure numbers.

Different disciplines may choose to handle the appraisal of the scores differently. For simplicity sake, rather than because it is more correct, I will presume raters will choose to add the scores together within categories to obtain cumulative scores for inputs, for throughputs, and outputs/outcomes. Finally, they add the cumulative scores for throughputs and outputs/outcomes to arrive at the project's overall results score (Throughputs + Outputs/Outcomes).

To assess goal attainment, the evaluators compare the results score of the research as reported to the maximum possible results score had all throughputs and outputs/outcomes been present or scored at their highest level.

Alternatively, if a modification of prior probability approach is employed, once the experts determine the prior probability of a particular study result, then they can calculate the posterior probability of the result from the study they are appraising.

Appraising research from the standpoint of whether and to what degree it modifies a prior probability is an approach especially suited to appraising theoretically guided research. Theories make predictions of particular results and in principle it should be possible based on a theory for experts in the field to assign a prior probability and then look at what the study did. The modification of a prior probability approach allows value for replications that need to be done when results of great scientific or policy significance are involved. Note while replications would modify a new prior resulting from the initial investigation's modification of a prior probability that might have been exclusively derived by fiat, it gets some credit, albeit not as much as the first study, for its contribution.

To assess efficiency, find the ratio of the results score to the cumulative score for inputs. The more inputs that are required to produce results, the less efficient is the research process. Maximum efficiency is represented by the ratio of the maximum expected results score to one since all projects require some inputs (R:1). The larger the ratio of results to inputs the more efficient the project or initiative.

In order to control for rater error/bias, ideally reports and publications should be *independently* scored by a number of raters-stakeholders, that is, without their consulting with one another prior to scoring the variables. Inter rater agreement can then be assessed with familiar statistical techniques such as with Kendall's *W* of concordance (see Solomon Diamond *1959)* to see how closely the raters come to a similar conclusion. Depending on the protocol chosen, variable specific findings that appear to be outliers may be eliminated before finding the average scores for a particular variable.

Situations can arise where the raters wish to consult with one another prior to scoring the study. This would occur when externalities are significant but different stakeholders are affected differently by the project as it unfolds. Because externalities can dwarf in importance the other outcomes and results, sometimes the only real issue in appraising the project is the accurate assessment of externalities (See Edwards and Newman 1982:62-64).

In order to assess efficiency, reviewers must have sufficient information about the researchers' access and utilization of resources to identify and score input variables. If this information is not available from the publications, it may have to be solicited from the author or authors. The person soliciting the information would be someone other than the reviewers so that the reviewers may remain blind to the identity of the author(s).

A note on externalities

The analysis has proceeded on the assumption that consideration of externalities is irrelevant. Indeed, in much routine research in science that assumption is justified. For instance, macro-biologists may investigate the possibility of new insect species in an ecosystem that has not been well studied. They would visit the area and conduct studies designed to trap and identify insects.

However, sometimes this assumption about externalities (or "spillover effects" as Mishan prefers to term them) is not warranted. If externalities exist, then certainly the evaluation of a research project can benefit from the insights of economists. In the example just given, while the survey may not have externalities, subsequent research, e.g., close study of a newly discovered insect species, could have such spillover effects. For example, the scientists may notice that the insects do not become infected with bacteria when they ingest blood from a sick animal and this insight could yield clues to a new class of drugs for fighting bacterial or viral infection, or have some other promising medical application.

In the STEM fields, especially biomedical sciences and applied physics, externalities frequently arise from research. To give another example, an engineering study resulting in superior cameras for use in a satellite monitoring a potential adversary's military installations also may turn out to be useful for archeological studies for which it may never have been intended. Including these externalities when they are known is essential in assessing a project's worth, and the scientific contributions of the project's personnel.

Various kinds of externalities are distinguished in the literature and only a limited discussion of them is possible here. It is important to point to "real externalities", which can be either negative or positive, as one effect of research. For instance, new findings from applied research on computer printing processes might allow a manufacturer to produce faster computer printers and/or multi-color printers that print in truer colors. The real externalities in this case are easily measured; the difference in price per unit multiplied by the number of units sold compared to the previous model since we expect the better computer printers can be offered at a higher price than the older models. If the number of computer printers sold also increases, the additional sales are relevant as well to assessing the real externality of the applied research.

Research can also have other kinds of externalities. Economists, for example, distinguish pecuniary externalities from non-pecuniary externalities. The former would affect the prices paid for goods already available. For instance, a pecuniary externality of the new products would be to lower the price of the other slower printers (or machines without the same color printing capabilities of the new machine). By forcing competitors into lowering their prices, more people (including people who could not afford the older technology) would be able to buy a machine. This is an example of a positive externality of the new machine—although other (competitor) manufacturers might not see the new situation as a positive development (because the loss of revenue to them in the short term is a real [negative] externality).

In principle, externalities can be measured in cardinal terms (e.g. USD). However, as Arluck (2006) emphasized in a personal communication, "Externalities are notoriously difficult to measure, as are their associated social costs and social benefits." He added, "The absence of property rights, and the absence of markets for these rights requires estimation of implied social costs and benefits."

Because many important decisions about what research to pursue depend on appraisals of the externalities of particular studies compared to other research (or even of non-research uses of the money), good estimates of the social costs and benefits of the research are a necessary aspect of the appraisal process. Consider the following hypothetical example. Researchers at Columbia University have developed an experimental ape with Alzheimer's-like brain plaques as a prelude to developing new chemotherapies. Initially, researchers believed these experimental apes will be immensely valuable because the disease process they are investigating with the apes affects large numbers of individuals. However, controversy exists as to whether the plaques are causative or not or even unrelated to Alzheimer's disease. Further research performed with the apes suggests the value of these apes for Alzheimer's research is much less than initially supposed. This disappointing news, however, is partially offset by the enthusiasm of other scientists who indicate the apes have great value for studying a number of rare brain illnesses even if their value in Alzheimer's disease is unclear at best.

At what point in the above example should the externalities be measured? Clearly the answers given by an externalities assessment done initially would be different from those obtained subsequently, possibly on the order of millions of dollars or more.

This example does not even take into account the issue of what is the value of individual researcher contributions in the ape research. Suppose one of the principal investigators is being evaluated for tenure. A tenure decision made immediately after the experimental animal has been developed might have a different outcome than one made after research shows experimental animals to not be especially useful in Alzheimer's research. On the other hand, once it is shown the animals are valuable in studying other rare brain disorders, the tenure decision may have a different outcome again. Tenure decisions, of course, are usually made at a fixed point in time and the university does not have the luxury of waiting for the external value of the research to become evident before deciding whether to grant tenure.

Economists do not concern themselves with externalities that involve the transfer of some benefit from one party to another, that is, a loss for one party and a gain for another. This, however, may be relevant in the assessment of a project from the perspective of the stakeholders.

Example: Assessing goal attainment where goal attainment EQ results

(Innovative Throughputs + Outputs / Outcomes)

A worked out example using hypothetical data is shown below. Note in this example the scores presented are assumed to be final project scores after initial individual raters scores having been tested for inter-rater agreement. Also note in the worked out example below externalities are *not* computed. When externalities are known to be present, they should be computed before the other dimensions. In the unusual situations when the externalities are extremely high, it is unnecessary to compute anything else.

Value of Variable

Significance to future research
Low (1) Moderate (2) High (3)

Innovative throughputs

Creates a new data base	2
Examines a new population	2
Uses cross-disciplinary theories to validate	1
Score: 5 out of possible	9

Outputs/Outcomes

Replicates previous finding	1
Provides new instrument	3

Externalities (if known)	0 (See note below)

Score:	4 out of possible	6
Results:	9 out of possible	15
Goal Attainment Score:		met 60% of the maximum expectation

Value of Variable
Contribution to completion of research
Low (1) Moderate (2) High (3)

Inputs

Funding	1
Equipment/research facilities	1
Pre-existing question	3
Collaborators/research assistance	3

Score:	8 out of possible	12

Efficiency ratio: 9: 8 (1.12)
Maximum efficiency ratio: 15: 4 (=3.75)
Lowest efficiency ratio: 5:12 (=0.41)
Efficiency score: 29% (1.12/3.75=0.298) of maximum expectations

Assessing the productivity of multiple projects and initiatives

Reviewers assess the productivity of multiple projects and initiatives conducted by multiple researchers as follows:

(1) Ensure that all research projects/initiatives under review pertain to the same topic or area of study.
(2) Review the reports/publications related to each project/initiative individually, in date order, and against the same set of weighted variables.
(3) Calculate the goal attainment and efficiency of each project/initiative and compare.

As with single projects/initiatives, an innovative throughput, output, or outcome is counted only once, the first time it is reported or recognized regardless of how many times it may be referenced in subsequent reports or publications. The date that such information is first reported for each project/initiative is entered. If upon completion of the review of all the material submitted for all the projects/initiatives, there are several projects/initiatives that report the same innovation, the project/initiative for which the information was first reported is credited with innovation.

As compared to the current model for assessing research productivity, the new measure does the following:

(1) it evaluates and compares *research* rather than the researcher;
(2) it provides a uniform set of weighted, quantifiable variables against which to assess the quality of one research project or to compare the quality of multiple research projects within a the same research area;
(3) it provides the means to digitize and compare past evaluations of research in a particular subject area including the means of determining the impact on findings of changes created in specific variables;
(4) it permits researcher blind assessment of research without consideration of personal attributes;
(5) if desired, it permits a means to compare the research of individual research institutions or researchers which is free of personal attributives. Provides a quantitative basis for assessing the availability of resources and its influence on the conduct and outcome of the research;
(6) it reflects the relationship of organizational factors on the availability of resources and consequently on the outcome of research;
(7) it distinguishes between the content of individual publications; and
(8) it distinguishes between publications providing significant new information and those not doing this.

Discussion

This paper argues for, and presents, a new approach to scientific productivity assessment, based on the non-experimental program evaluation literature and the public administration tradition of project evaluation. The new model corrects many of the weaknesses of the simple quantity measurement of performance, and when properly used, is a superior approach to personnel evaluation.

As with any tool, however, it is not perfect. A disadvantage of the new measurement is its application is much more labor intensive. However, since the sets of variables can be digitalized and adapted for future use, the amount of labor needed to apply the model may decrease upon subsequent use.

Another limitation is the weighted variables for the model are selected and scored by experts in the subject area who may continue to harbor bias. Such bias may be mitigated by the extent to which the process is made transparent. Researchers can also help themselves in the evaluation process by emphasizing the need for the evaluation process to consider externalities when these clearly are beneficial.

Despite the additional work involved in utilizing the new measure, as a tool for expanding opportunities for women and minority candidates for tenure and promotion, it may be justified since it provides a uniform basis of comparison that can, if necessary, be validated via replication by another group of reviewers.

Endnotes to Appendix One

1. This Appendix is a revision of a poster presentation originally prepared for the November 2005 meetings of the Association for Public Policy and Management in Washington, D.C. Dr. Carmela Triolo Della Porta prepared the poster presentation with me for that meeting; however, she in no way is responsible for the present version.

Appendix Two

Data comparing countries in respect to women researchers

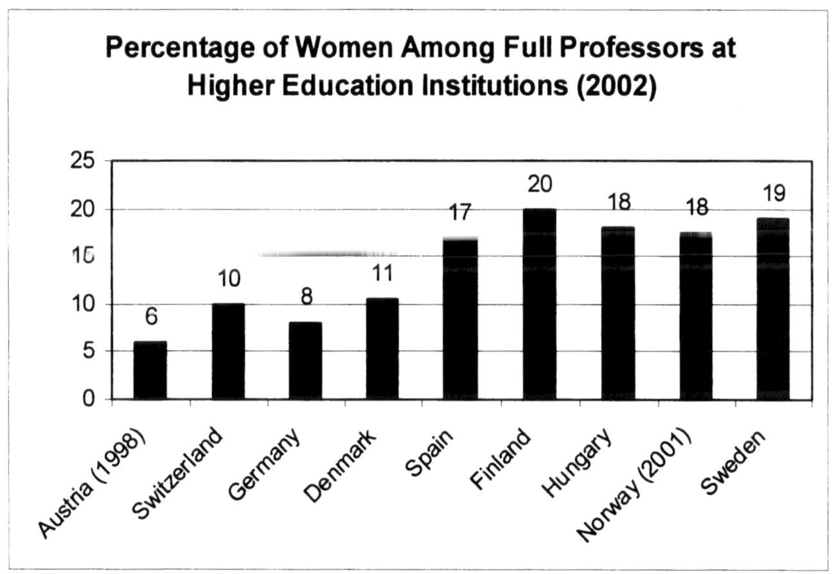

Figure A1: Percentage of Women among Full Professors at Higher Education Institutions by Country

Source: Langberg, Kamma (2005) "Gender Gap and Pipeline Metaphor in the Public Research Sector", OECD Workshop on Women in Science.

Table A1: Number of Female Researchers and their Percentage of All Researchers in Japan, 1992-2004

Year	Number (1,000s)	Percentage (%) of All Researchers
1992	49.2	7.9
1993	53.6	8.3
1994	57.2	8.6
1995	61.1	8.9
1996	64.9	9.3
1997	70.5	9.8
1998	74.2	10.2
1999	76.1	10.1
2000	80.7	10.6
2001	82.0	10.9
2002	85.2	10.7
2003	88.7	11.2
2004	96.1	11.6

Source: Ogawa, Mariko (2005) "The Present Condition and Problems of Women in Science and Technology in Japan", presented at the OECD Workshop on Women in Science, 16/17 November 2005.

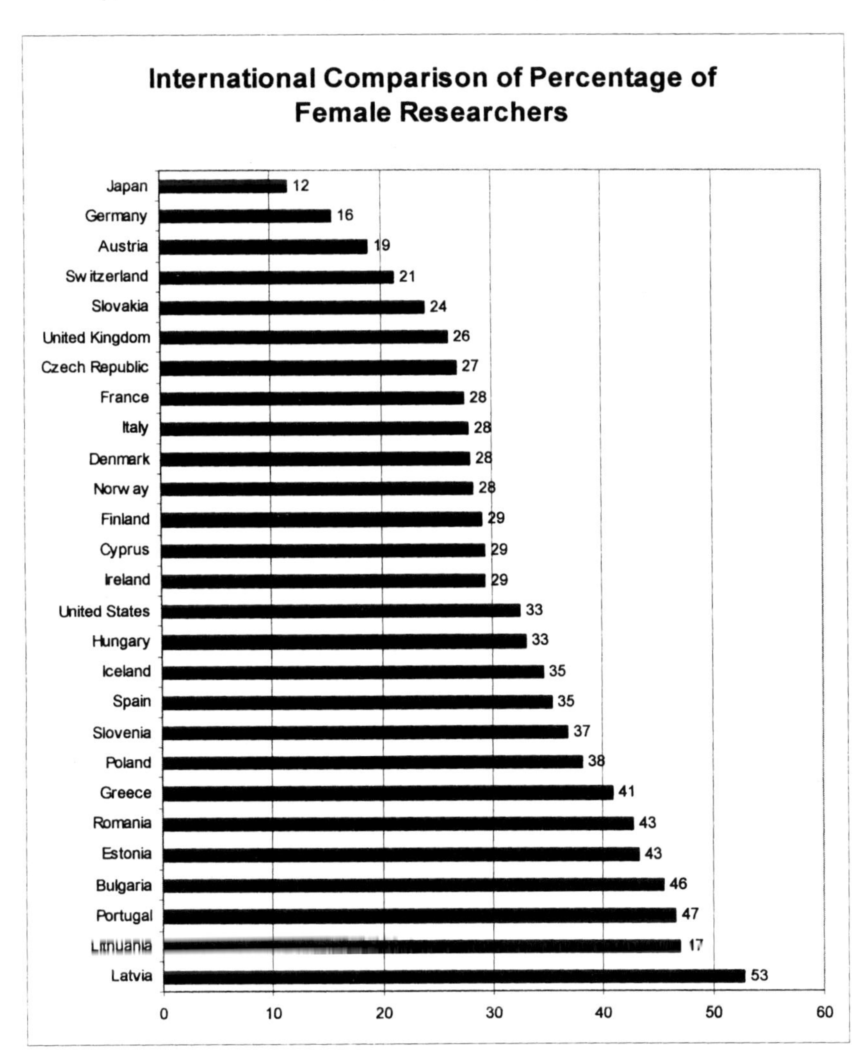

Figure A2: International Comparison of Percentage of Female Researchers
Source: Ogawa, Mariko (2005) "The Present Condition and Problems of Women in Science and Technology in Japan", presented at the OECD Workshop on Women in Science, 16/17 November 2005.

Bibliography

Arluck, Gregory
2006 "Personal Communication".

Association for Women in Science (AWIS)
2005 "Women in Science", Conference Report of the Association for Women in Science National Conference on Women in Science, Technology, Engineering, and Mathematics Disciplines, 23-24 June 2005, Smith College, Northhampton, MA.

Astin, Helen
1969 *The Woman Doctorate in America*, NY: Russell Sage Foundation.

Balk, Walter L.
1990 "Organization Theories as Instigators of Public Productivity Improvement Action", A pre-publication chapter prepared for *The Public Productivity Handbook*. 1st ed. NY: Marcel Decker.

Begley, Sharon
2006 "He, Once a She, Offers Own View on Science Spat", *Wall Street Journal*, July 13, 2006.

Bombardieri, Marcella
2005 "Harvard Ams to Spur Advancement of Women," *The Boston Globe*, Feb. 4, 2005.

Bozeman, Barry and Elizabeth Corley
2004 "Scientists' Collaboration Strategies: Implications for Scientific and Technical Human Capital" in *Research Policy,* available at *www.rvm. gatech.edu.*

Bush, Vannevar
1945 *Science: The Endless Frontier,* U.S. Govt Printing Office. Washington, D.C., available at *www.nsf.gov/od/lpa/nsf50/vbush1945.htm.*

Carroll, Felix
2005 "A Field of Equals: A Growing Number of Girls Dispel the Belief that Females Aren't Interested in Science", Albany (NY) *Times-Union,* Feb. 1, 2005.

Cockburn, Cynthia
1985 Machinery of Dominance: Women, Men and Technical Know-How, London, England: Pluto, quoted in Wajcman (1995).

Cohen, M.D., J.G. March, and J.P. Olsen.
1972 "A Garbage Can Model of Organizational Choice", in *Administrative Science Quarterly*, 17:1-25.

Cole, Jonathan R.
1981 "Women in Science", *American Scientist 69 (July-August)*:385-391 cited in Fox (1992).

Cole, Jonathan R. and Burton Singer
1992 "A Theory of Limited Differences: Explaining the Productivity Puzzle in Science", in Zuckerman, H., J.R.Cole, and J.T. Bruer (eds.), *The Outer Circle: Women in the Scientific Community* (New Haven: CT: Yale Paperbound. originally1991, pp. 277-310, 319-323, 338-340).

Cole, Jonathan R. and Harriet Zuckerman
1984 "The Productivity Puzzle: Persistence and Change in Patterns of Publication of Men and Women Scientists", in *Advances in Motivation and Achievement*, P. Maehr and M. W. Steinkamp, eds., pp. 217-56, Greenwich, CT.: JAI Press quoted in Valian, 1998.

Cook, Thomas D. and Donald T. Campbell
1979 *Quasi-Experimentation Design and Analysis Issues for Field Settings* (Boston, MA: Houghton Mifflin Company).

Crane, Diana
1977 "Social Structure in a Group of Scientists: A Test of the 'Invisible College' Hypothesis", in *American Sociological Review*, 34:335-352.

Dean, Cornelia
2005 "For Some Girls, the Problem with Math is that They're Good at It", NY *Times* Feb. 1, 2005).

Della Porta, Carmela Triolo and Robert Leslie Fisher
2005 "Assessing Research Productivity with a Gender Neutral Model", Poster Presentation at Association for Policy Planning Research and Management (APPAM), Washington D.C.: November 2005.

Diamond, Solomon
1959 *Information and Error,* New York: Basic Books.

Dietz, James
2004 "Scientists and Engineers in Academic Research Centers-an Examination of Career Patterns and Productivity", Ph.D. diss. Georgia Institute of Technology.

Dix, L.
1987 *Women: Their Under Representation and Career Differentials in Science and Engineering*, Washington D.C.: NAS Press.

Edwards, Ward, Marcia Guttentag, and Kurt Snapper.
1975 "A Decision Theoretic Approach to Evaluation *Research", in Handbook of Evaluation Research:* Beverly Hills, CA: Sage Pubns.

Edwards, Ward and Robert Newman
1982 *Multiattribute Evaluation:* Sage University Paper Series on Quantitative Applications in the Social Sciences, 07-026, Beverly Hills, CA: Sage Pubns.

Elden, Max and Sanders, Steven
1996 "Vision Based Diagnosis, Mobilizing for Transformation Change in Public Organizations", *Public Productivity and Management Review*, 20(1):11-23.

Epstein, Cynthia Fuchs
1970 *Woman's Place: Options and Limits in Professional Careers*, Berkeley: University of California Press.

Etzkowitz, Henry, Carol Kemelgor, and Brian Uzzi
2000 *Athena Unbound:The Advancement of Women in Science and Technology,* Cambridge England: Cambridge University Press.

Feldt, Barbara.
1995 *The faculty cohort study: School of Medicine,* Ann Arbor, MI: Office of Affirmative Action, cited in Fox, Mary Frank (1991).

Finkelstein, M.
1998 "The Status of Academic Women: An Assessment of Five Competing Explanations", in *Review of Higher Education,* 7:223-246, quoted in Valian (1998).

Feldt, Barbara
1986 *The Faculty Cohort Study: School of Medicine,* Ann Arbor, MI: Office of Affirmative Action cited in Fox, Mary Frank (1995).

Fisher, Robert Leslie
2005 The Research Productivity of Scientists How Gender, Organization Culture, and the Problem Choice Process Influence the Productivity of Scientists, (Lanham, MD: University Press of America).

Fodor, Kate
2005 "Women on the Rise", available at *www.the-scientist.com/2005/11/07/ s/411.*

Fox, Mary Frank
1992 "Gender, Environmental Milieu, and Productivity in Science, in Zuckerman, H., J. R. Cole and J. T. Bruer (eds.), *The Outer Circle: Women in the Scientific Community,* (New Haven: CT: Yale Paperbound. Originally 1991) pp 188-204,314,331-334.

____
1995 "Women and Scientific Careers", in Jasanoff, Sheila and Gerald E. Markle, James C. Petersen, Trevor Pinch (eds.) *Handbook of Science and Technology Studies,* (Thousand Oaks, CA: Sage Publications), pp.205-223.

Gordon, Henry A.
1990 *Who Majors in Science? College Graduates in Science, Engineering, or Mathematics from the High School Class of 1980,* (NCES 90-658), Washington, D.C.: U.S. Government Printing Office.

Handelsman, Jo, N. Cantor, M.Carnes, D. Denton, E. Fine, B. Grosz, V. Hinshaw, C. Marrett, S. Rosser, D. Shalala, and J. Sheridan
2005 "More Women in Science" in *Science,* 309, pp. 1190-1191.

Hartline, B.K. and D. Li, eds.
2002 *Women in Physics: The IUPAP International Conference on Women in Physics*, American Institute of Physics.

Hartline, B.K. and D. Li, eds.
2005 Women in Physics: The 2nd IUPAP International Conference on Women in Physics, American Institute of Physics.

Hermann, Claudine
2002 "The European Commission Report on Women and Science and One Frenchwoman's Experience", in Hartline, B.K. and D. Li, eds. *Women in Physics: The IUPAP International Conference on Women in Physics*, American Institute of Physics.

Holcombe, Randall G. and Russell S. Sobel
2000 "Consumption Externalities and Economic Welfare", in *Eastern Economic Journal,* p. 4,8.

Hornig, Lillie (editor)
2003 *Equal Rites, Unequal Outcomes,* (NY: Kluwer Academic/Plenum Publishers).

Ivey, Elizabeth
2005 in "Women in Science", Conference Report of the Association for Women in Science National Conference on Women in Science, Technology, Engineering, and Mathematics Disciplines, 23-24 June 2005, Smith College, Northhampton, MA.

Jasanoff, Sheila, Gerald E. Markle, James C. Petersen, and Trevor Pinch (eds.)
1995 *Handbook of Science and Technology Studies,* (Thousand Oaks, CA: Sage Publications).

Kodate, Kashiko and Eiko Torikai
2005 "Promoting Gender Equality in Physics in Japan", in *Women in Physics: The 2nd IUPAP International Conference on Women in Physics*, Hartline, B.K. and D. Li, eds., American Institute of Physics.

Kolpin, V. W. and L. D. Singell, Jr.
1996 "The Gender Composition and Scholarly Performance of Economics Departments: A Test for Employment Discrimination ", in *Industrial and Labor Relations Review,* 49: 408-23.

Konehamp, Barbel, Beate Kravis, Martina Erlemann, and Corinna Kausch
2001 "Survey of Male and Female Opportunities for Men and Women in Physics", available at www.dpg-physik.de/static/fachlich/akc/Studie-Bericht-engl.pdf.

Lackland, Ann Childers
2001 "Students Choices of College Majors that are Gender Traditional and Nontraditional", Journal of College Student Development, Jan/Feb.

Liebowitz, S. J. and Stephen E. Margolis
1996 "What are Network Effects?", available at *www.utdallas.edu/~liebowit/ palgrave/ network.html.*

Long, J. S.
1992 "Measures of Sex Differences in Scientific Productivity", *Social Forces* 71:159-178 cited in Valian, Virginia, *Why So Slow?*

March, James G.
1988 *Decisions and Organizations, (*Oxford, England, Basil Blackwell Ltd.).

Merton, Robert K.
1973 *The Sociology of Science* (Chicago: University of Chicago Press).

Molinari, Elisa, Maria Grazia Betti, Annalisa Bonfiglio, Anna Grazia Mignani and Maria Luigia Paciello
2002 *" Women in Physic in Italy: The Leaky Pipleline"*, available at *www.infn. it/cpo/contributi/contributi/2pages_elisa.pdf.*

National Academy of Sciences
1991 *Women in Science and Engineering: Increasing their Numbers in the 1990s: A Statement on Policy and Strategy.*

National Academy of Sciences
2005 *Graduate Enrollment in Science and Engineering Programs Up in 2003, But Declines for First-Time Foreign Students,* (prepared by Julia Oliver), *www.nsf.gov/statistics/inbrief/nsf05317.*

Parsons, Keith (ed.)
2003 *The Science Wars, (*Amherst NY: Prometheus Books).

Parsons, Talcott
1951 *Social System, (*NY: The Free Press).

Pinker, Steven and Elizabeth S. Spelke
2005 *The Science of Gender and Science: Pinker vs. Spelke-A Debate* (An Edge Special Event, May 10, 2005), available at *www.edge.org/3rd_culture/debate05 /debate05_index.html.*

Pollard, William
1986 *Bayesian Statistics for Evaluation Research,* Beverly Hills, CA: Sage Publns.

Pressman, Jeffrey and Wildavsky
1973 *Implementation:How Great Expectations in Washington are Dashed in Oakland; or Why Its Amazing Federal Programs Work at All, This Being a Saga, (The Oakland Project Series).* Berkeley, CA: University of California Press.

Rohrbaugh, John
1981 "Operationalizing the Competing Values Approach", Public Productivity Review, 5:141-159.

Rosen, Ellen Doree
1993 *Improving Public Sector Productivity Concepts and Practice* (Newbury Park: Sage Publications).

Rosser, Sue V.
2004 *The Science Glass Ceiling,* Cambridge, England: Cambridge University Press.

Sachs, Albie and Wilson, Joan Hoff
1978 *Sexism and the Law: Male Beliefs and Legal Bias in Britain and the United States,* New York: Free Press.

Sandow, Barbara and Corinna Kausch
2005 "Women in Physics in Germany" in *Women in Physics: The 2nd IUPAP International Conference on Women in Physics*, Hartline, B.K. and D. Li, eds., American Institute of Physics.

Senge, Peter M.
1990 *The Fifth Discipline: The Art & Practice of the Learning organization,* NY: Doubleday/Currency.

Sonnert, Gerhard
1995 "What makes a good scientist? Determinants of peer evaluation among biologists" *Social Studies of Science,* 25:35-55, cited in Valian, Virginia, *Why So Slow?*

Stack, S.
1994 "The effects of gender on publishing: The case of sociology" *Sociological Focus,* 27:81-83, cited in Valian, Virginia, *Why So Slow?*

Subowa, W.
2002 "Indonesian Women Physicists" in Hartline, B.K. and D. Li, eds. *Women in Physics: The IUPAP International Conference on Women in Physics*, American Institute of Physics.

Sulloway, Frank
1996 *Born to Rebel (*NY: Pantheon Books).

Thompson. J. D.
1967 *Organizations in Action*, NY: McGraw.

Valian, Virginia
1998 *Why So Slow? The Advancement of Women,* (Cambridge: MA and London, England: MIT Press).

Van de Ven, A. H. and Joyce Wif
1981 P*erspectives on Organization Design and Behavior,* (NY: John Wiley & Sons).

Van de Ven, A. H.
1986 "Central Problems in the Management of Innovation", *Management Science*, 32 (5), 590-607.

Van de Ven, A. H. and Koenig, Richard Jr.
1976 "A Process Model for Program Planning and Evaluation", *Tour of Economy and Business*, 161-170.

Wajcman, Judy
1995 "Feminist Theories of Technology" in Jasanoff, Sheila, Gerald E. Markle, James C. Petersen, and Trevor Pinch (eds.), *Handbook of Science and Technology Studies* (Thousand Oaks, CA: Sage Publications, pp.189-204).

Wenneras, C. and A. Wold
1997 "Nepotism and Sexism in Peer Review", *Nature* 347: 341, cited in Hermann, Claudine.

Xie, Yu and Kimberly A. Shauman
1998 "Sex Differences in Research Productivity: New Evidence About An Old Puzzle", *American Sociological Review,* vol. 63 (December: 847-870).

____
2003 *Women in Science* (Cambridge MA: Harvard University Press).

Zeldenrust, Sjerp
1990. *Ambiguity, Choice, and Control in Research,* Universiteit van Amsterdam, Netherlands, Ph. D. diss.

Zuckerman, Harriet
1987 "Persistence and Change in the Careers of Men and Women Scientists and Engineers: A Review of Current Research" in *Women: Their Underrepresentation and Career Differentials in Science and Engineering*, L. S. Dixon, ed., pp. 123-156, Washington, D.C.: National Technical Information Service, cited in Valian, Virginia, *Why So Slow?*

———

1992 "The Careers of Men and Women Scientists: A Review of Current Research", in Zuckerman, H., Cole, J. and Bruer, J. T. (eds.), *The Outer Circle: Women in the Scientific Community* (New Haven, CT: Yale Paperbound. Originally 1991).

Index

About the author

The author is a product of the New York City public school system including Peter Stuyvesant High School, a specialized high school for students with an interest in science and engineering, from which he graduated in 1963. At that time the school was for male students only. He attended City College of New York from which he graduated in 1967, majoring in sociology. Besides graduating *cum laude*, he received General Honors and the Ward Medal in Sociology. From 1967-1969 he studied sociology at Columbia University receiving his Master of Philosophy in 1976. In 1972 Mr. Fisher joined the New York State Division of Probation as a research analyst and embarked on a career spanning more than three decades in state government as a criminal justice planner and evaluator.

Now retired, Mr. Fisher is an independent scholar with interests in organization analysis and program evaluation. He is the author of a previous book in the sociology of science, *The Research Productivity of Scientists,* also published by University Press of America (2005). He is currently working on a study of how changes in the engineering professions and their culture are impacting the penetration of women in these disciplines and also on a novel about Peace Corps volunteers in Africa.